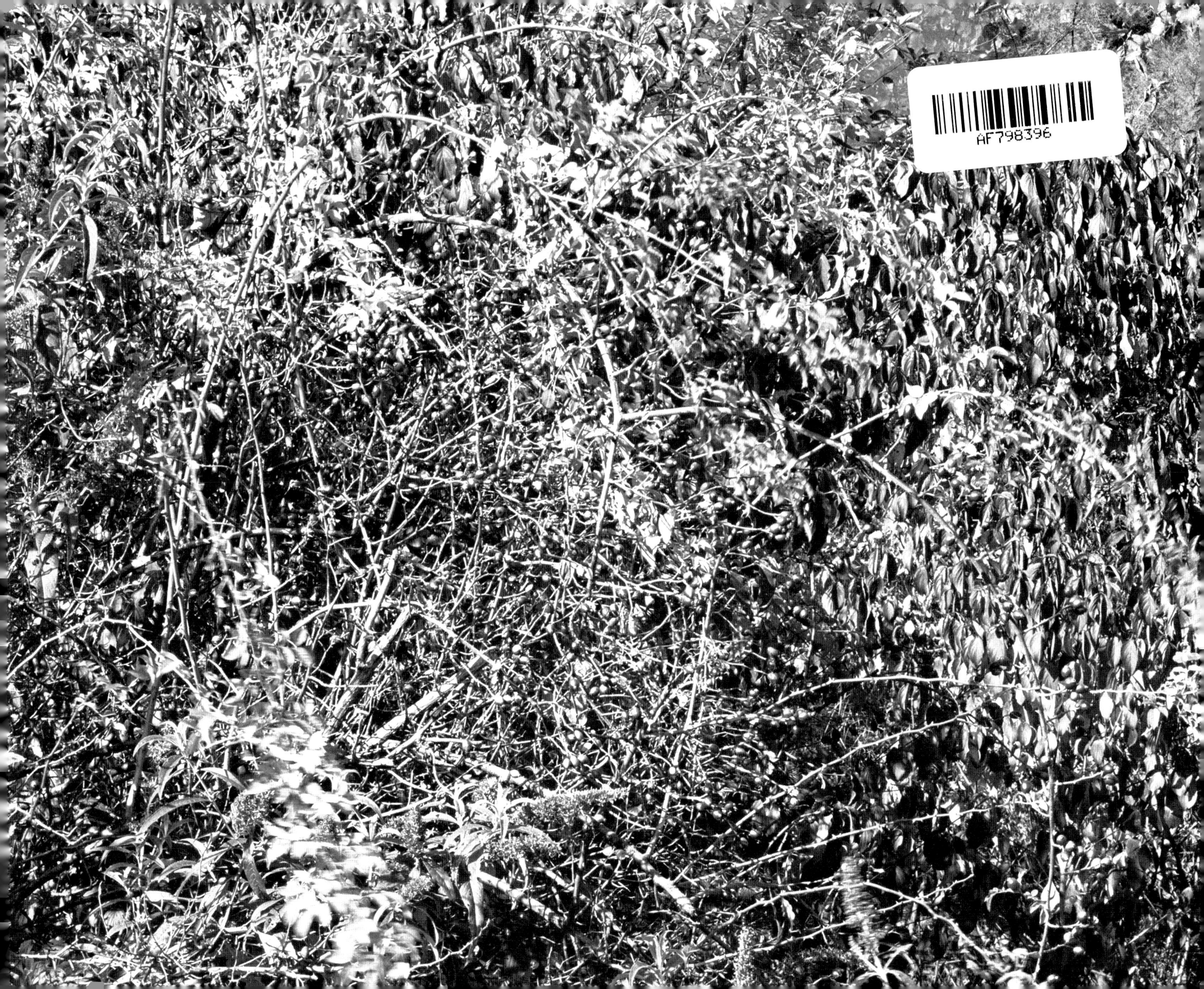

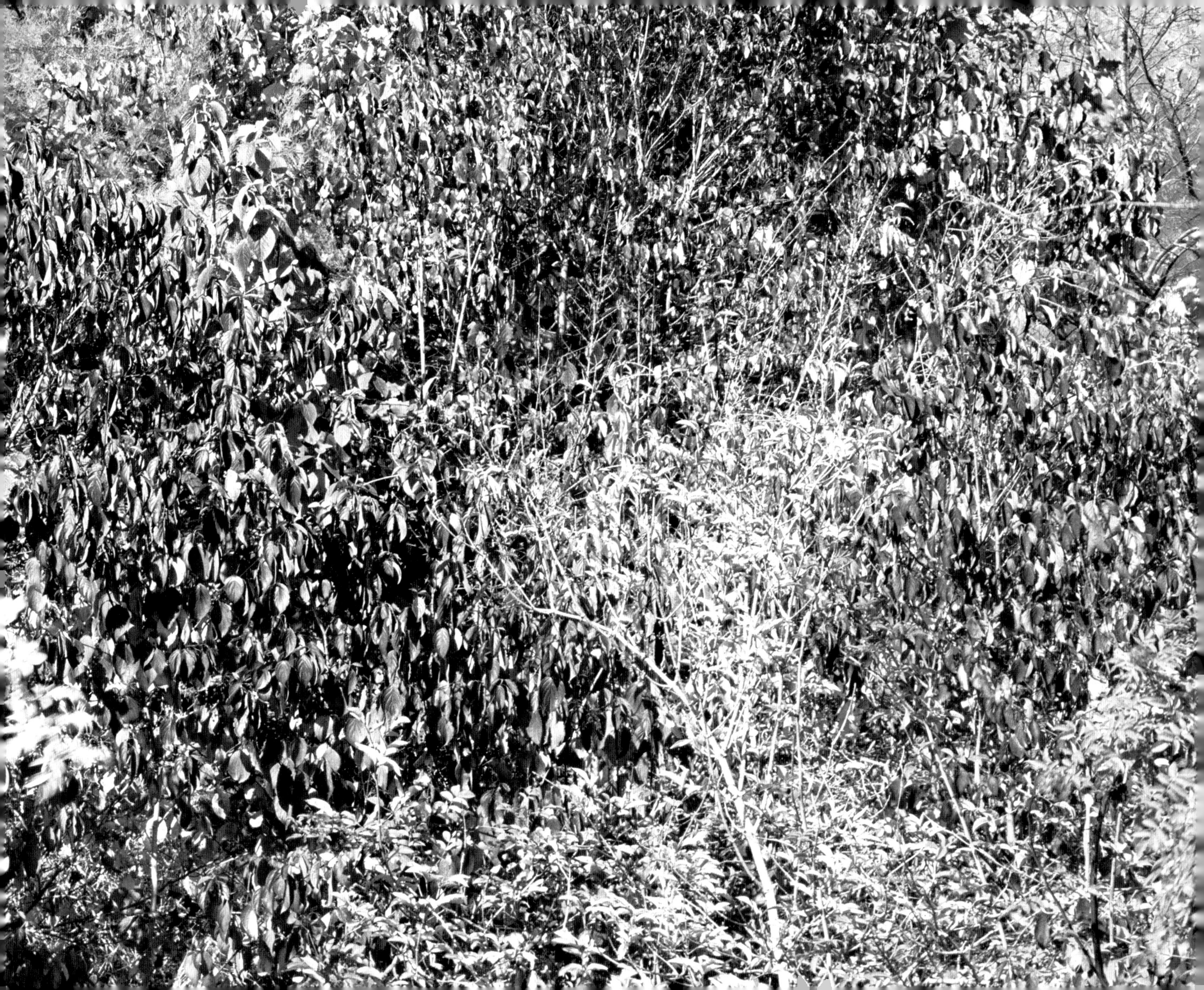

Landscape A

Nicolas Faure

Published on the occasion of the exhibition
Nicolas Faure, Landscape A

Musée de l'Elysée, Lausanne
17 November 2005 – 5 February 2006

This book and accompanying exhibition has been made possible
with the support of the Swiss Federal Office of Culture, Bern
and the Swiss Federal Roads Authority, Bern.

A1
A1a
A1.1
A2
A3
A4
A5
A6
A7
A8
A9
A12
A13
A14
A16
A20
A21
A28
A50
A51

To my parents Yvette and André,
my wife Sophie,
and my children Alice et Mathias

Landscape through the Art of Seeing

Daniel Girardin, curator, Musée de l'Elysée, Lausanne

I think landscape into being, I am its conscience. Cézanne

Since the early '90s, Nicolas Faure has created a vast body of landscape photographs of contemporary Switzerland. Faure's images capture the signs that identify intersections of the traditional and the modern. These signs are sometimes obvious, but they can also be tenuous, subtle, or even unnoticeable without the help of the camera. Faure has been pursuing this line of enquiry in successive photographic projects for fifteen years. The series *Paysages A* (or "A" Landscapes, after the Swiss national highways) represents the culmination of his efforts.] [In all, Faure has produced three landscape series. They share the same spirit and are absolutely coherent in terms of meaning. The first shows isolated rocks dotting—apparently erratically—contemporary landscapes (1991); the second, highways (2000); and the third, the "A" landscapes (2005). There is an evolution here: from a reading of landscape in conventional symbolic terms, towards something entirely new. Thus Faure paves the way to a new reading of the contemporary environment. In all his work, the photographer casts a contemporary light on signs of the traditional be they natural, mythic, or cultural.] [Single large stones preserved as traces of nature untrammeled, and carefully situated in various places on Swiss territory, are larded with symbolism, allowing the photographer to present his ideological landscapes with a lightly critical sense of humor. In the same vein, *Autoland*[1)], Faure's second series, is like a vast biography in images of Swiss highways. It explores what are quite literally throughways, or non-places, unrelated to their surroundings and without history or memory. This is a characteristic that highways share with airports, industrial zones, high-speed train lines, and shopping malls[2)], to name the most obvious related domains. By their very nature minimalist and utilitarian, highways are disembodied transit areas.] [*Paysages A*, Faure's third series, focuses not on the highways themselves but on the landscaping flanking them. This landscaping is part of the large-scale integration of highways into environments they radically transform. Through patient, cosmetic, artificial and idealized gardening, the aim is to create and maintain green zones, where plants and shrubs grow, thus constituting a screen between the no man's land of the highway and the territory it passes through. The freshly minted landscapes do, however, strive to make reference to nature, which is to say to a form of identity. Faure's images of these highly varied universes add up to a powerful overview of Swiss highway landscapes, which are very different from those of other European countries where there is less fast-changing scenic diversity and systems are not necessarily maintained by public works departments.] [The Utopic, illusion-inducing artificial gardens that Nicolas Faure has discovered along Switzerland's highways can only be seen from a distance. They are invisible to those in passing vehicles because of the speed at which the vehicles are traveling. Nor do the gardens reveal themselves to pedestrians—the spaces are for the most part enclosed and in any case off-bounds. These "compensation zones", created to diminish the effects of the widespread use of concrete, usually border on the highways directly. Faure's photographs make it possible not only to see differently; they take a new look at landscape, showing a form of beauty in no way nostalgic and yet central to modern concerns, including ecological ones. His images also make it possible to understand the bond that man—or, more prosaically, the federal administration—has to nature, or at least to the idea of it.

Depicting landscape—a story of images

The concept of landscape is vast, with a history going back to 1549. Landscape is a complex cultural construct, a perception of place that exists only when it is represented or described as such in imagery or literature. For a long time painting took an essentially esthetic approach to landscape, until the Age of Enlightenment, in the 18th century, when the link was made between it and nature.] [Photography, after its appearance in the 19th century, became a major landscape genre. The medium made it possible for scientific or documentary prerogatives to coexist with esthetic ones. From its invention, photography has oscillated between art and document, depending on the importance given to details or, conversely, the accent placed on interpretation. Many landscape photographs viewed as strictly documentary in the 19th century were, a mere 50 years later, considered works of art—one of the miracles of passing time that also points to a change in the artistic and social criteria attaching to the way photographs were read.] [Because of this long history, contemporary landscape photography is viewed with a certain amount of distance. The documentary and the artistic make easy partners today; indeed, their complementarity is strong. Landscape is at once a subjective interpretation (the photographer's view), a plastic intent underpinned by the quality of the print (often large format), and a documentary fragment because photography, which is figurative by nature, makes it possible to capture a great number of signs.] [The landscape is at once the thing and its representation. It is the result of a transformation of place and that of the transformation brought about by the eye of an artist; these add up to models. Over the course of several centuries landscapes have been modeled and remodeled, first by the development of the countryside and then by industrialization. In the West, there is no more "natural" nature; it is all part of territorial management based in concepts not dissimilar to those of gardens and landscaping and that require images to become landscapes.

Landscape and environment

In Switzerland particularly, photography in the 19th century already challenged traditional landscapes even as it made it possible to integrate them into a collective vision. Switzerland's landscapes are part of its natural and cultural heritage. During the course of the 18th, 19th and 20th centuries its preservation, management and transformation led to many political measures[3)]. Notions of growth and unlimited progress brought about changes that seriously endangered the environment and in turn led to a change in consciousness in the '70s that was marked by the creation, in 1971, of the Swiss Agency for the Environment. The construction of the country's highway network, which began in the euphoric '50s, is coming to fruition in today's climate of economic downturn and heightened awareness of environment-linked issues including pollution of all kinds, not least from noise. The construction process, however, is irreversible.] [By the end of the 20th century, the operative idea in terms of industrial growth was achieving balance. Translated into landscape terms, this meant building transportation networks of extremely high quality accompanied by "natural spaces"—rather like tributes paid for covering to saturation point large parts of a small territory with concrete, or monuments of illusion and artifice to the memory of nature.] [These supposedly "natural" spaces are part of the emerging landscape of highways, bridges, viaducts, tunnels, walls, and sloping embankments. They hide highways—the aim is to "humanize" the disembodied character of highways by these simulations of natures—and veil surrounding areas in a country where scenic changes progress very rapidly from urban to industrial, lake to mountain, open air to tunnel.] [This diversity precludes any sensation of spatial immensity such as can be felt in the vast reaches of the United States, for example. The limitations also mean, however, an endlessly changing and surprising series of views, that Nicolas Faure has captured and transformed into plastic imagery of great beauty, and that are in themselves signs of Swiss identity.

Contemporary landscapes: the ruins of the future

These new territories, rich in both obvious and more subtle signs of Swiss identity, are transformed by photography into landscapes. Untamed spaces, without borders and chaotic, become images when they are framed, put in perspective, and given form. Faure had no critical intent when he embarked on this series. The only critical function at work here is that of photography itself. By showing a new way of seeing the contemporary, the photographer has gone from landscape permeated by history—the rocks in his first series—to a new kind of imagery, rootless and unexplored.] [The result is more formal and above all more pictorial, and corresponds to an idea of the unchangeable meaning of things. And therein lies the poetic paradox of contemporary landscape. It is also what makes it beautiful and, in terms of time, subjective. The viewer, as an active participant, brings his or her own cultural consciousness to the interpretation of images whose references are strictly contemporary, devoid of the picturesque or the nostalgic.] [Faure's landscape photographs—because they make evident what is otherwise seen only in passing, merely suggested or guessed at—make it possible to *see*. From the esthetic point of view, the images comprise an extremely coherent ensemble. Seen in terms of meaning, Nicolas Faure's landscapes stem from the desire to inscribe in the here and now—something that photography probably does best—at least some part of the heritage the 20th century has bequeathed future generations—and in which are secretly lodged the signs that will awaken the nostalgia of tomorrow.] [Straddling the fragile but clear border that separates art from the documentary, Nicolas Faure's photographic landscapes bring to mind the beauty and truth embodied by Renaissance ideals, and yet they do so from a resolutely contemporary perspective.

1) Cf. Nicolas Faure, *Autoland: Bilder aus der Schweiz, pictures from Switzerland*, Scalo, Zurich, 1998

2) Cf. André Augé, *Non-lieux, introduction à une anthropologie de la surmodernité*, La Librairie du XXIe siècle, Seuil, Paris, 1992

3) Cf. François Walter, *Les Suisses et leur environnement, une histoire du rapport à la nature du 18e siècle à nos jours*, Editions Zoé, Geneva, 1990

Marginal landscapes

Hans Ibelings, publisher of A10 new european architecture, Amsterdam

It is obvious to me as an outsider that the Swiss have a complicated relationship with the natural landscape. There are few places in Europe where nature is so awesome as here. Switzerland is conditioned by the immanent danger of avalanches, falling rocks and streams that may swell, in the wink of an eye, into raging torrents.] [Swiss society has the money, know-how and capacity to curb that danger, to canalize it and, to the extent possible with nature, to bring it under control. Love of nature in Switzerland goes hand in hand with a necessary, if occasionally somewhat obsessive, tendency to want to control that nature by technical and structural means. Dutch citizens, raised on the national mythology of the struggle against the water, understand this duality, even though the Alps will always remain to some degree beyond their comprehension.] [Of all the combinations of wild nature and civil engineering works in Switzerland, the motorway network ensemble is the most impressive. There are few places in the world where motorway and landscape are so stunningly interwoven. The combination of alpine landscape and modern motorway architecture has produced some sublime results. The bridges of Robert Maillart and the tunnel mouths of Rino Tami in Ticino are only a few of the most famous examples. And in many places in Switzerland the view from the road is similarly spectacular. Irrespective of circumstance or season, the mountains and lakes provide an unparalleled backdrop for a car journey.] [The postcard prettiness conceals the existence in Switzerland of yet another motoring landscape. A landscape that does not obtrude on the consciousness and that is often not even recognized as landscape. An artificial landscape created—and often maintained—with great care and precision and located in close proximity to the road. So close, in fact, as to be easily overlooked, this landscape of side and central reservations, of inner angles of motorway slip roads, of viaduct and noise barrier embankments.] [This motorway landscape, which does not fit the Swiss landscape cliché—lake in the foreground, snow-capped mountains in the background—has for some years been the subject of photographer Nicolas Faure. Asphalt and traffic rarely feature in his photographs. The overfamiliar Swiss landscape is usually absent, too, or is at best visible in the distance.] [What Faure captures in his images is the usually unremarked narrow margin between the road and what is normally regarded as landscape. This margin can—to borrow a term from Georges Perec—be labelled "infraordinary": it belongs to that category of objects and phenomena that are so ordinary as to escape our notice.] [Just as Perec sought to identify the infraordinary, so Faure has applied himself to photographing this twilight zone of the observable, where nothing remarkable seems to occur. Patient observation reveals it to be a landscape that is both familiar and alien. Familiar because we have all seen it so often without really looking at it, alien because in Faure's photographs it doesn't look the way you think.] [It is obvious that, contrary to the dominant trend in urban and landscape photography of recent years, Faure is not interested in aestheticizing or romanticizing the ordinary, or in evoking a sense of alienation. Without any explicit judgement or interpretation, he presents us with a reality. Ostensibly neutral, but not without a viewpoint, he shows us the result of the social habits of affluent, and consequently mobile, people.] [Faure's photographs seldom contain topographical distinguishing features. What we see is a landscape that could be anywhere or nowhere. This is an important message. Notwithstanding the many romantic ideas associated with being "on the road", the reality is that the motorist travels through an introverted, placeless zone that is not really part of the surrounding landscape. As such, it is a reflection and reinforcement of the condition of the motorist who, surrounded by car, airconditioning and music, remains aloof from the surroundings through which he or she travels.] [The anthropologist Marc Augé described the detached relationship between motorway motorist and the landscape in *Non-lieux; Introduction à une anthropologie de la surmodernité* (Éditions du Seuil, Paris 1992). Although he was referring to the French situation, his words have a much wider application: "France's well-designed autoroutes reveal landscapes somewhat reminiscent of aerial views, very different from the ones seen by travellers on the old national and departmental main roads. They represent, as it were, a change from intimist cinema to the big sky of Westerns." For Augé there is a crucial difference between the trunk road which runs *through* villages, towns and landscapes, and the autoroute which skims past villages, towns and landscapes. According to Augé, contact with the landscape from the motorway goes no further than the brown roadside signs alerting motorists to the places of interest they are passing: national monuments, noteworthy towns and villages, wine-growing areas, mountains, valleys. The autoroute motorist simply drives past them.] [Augé's observations about the distance between motorway and landscape accord with those of the celebrated architectural historian Reyner Banham as recorded in his classic study *Los Angeles: The Architecture of Four Ecologies* (Allen Lane The Penguin Press, London 1971). In it he describes his gradual realization that Los Angelenos do not regard the freeways that criss-cross Los Angeles as part of the city. "The first time I saw it happen nothing registered on my conscious mind, because it all seemed so natural—as the car in front turned down the off-ramp of the San Diego freeway, the girl beside the driver pulled down the sun-visor and used the mirror on the back of it to tidy her hair. Only when I had seen a couple more incidents of the kind did I catch their import: that coming off the freeway is coming in from outdoors. A domestic or sociable journey in Los Angeles does not end so much at the door of one's destination as at the off-ramp of the freeway, the mile or two of the ground-level streets counts as no more than the front drive of the house. In part this is a comment on the sheer vastness of the movement pattern of Los Angeles, but more than that it is a acknowledgement that the freeway system in its totality is now a single comprehensible place, a coherent state of mind, a complete way of life..."] [That notion of the motorway as a "single comprehensible place" is perhaps even stronger in Switzerland than in other countries. An often-heard Swiss complaint is that a lot of cross-border traffic takes advantage of the superb infrastructure without contributing anything to its upkeep. Many foreign motorists, motorcyclists, lorry drivers and bus passengers spend no more than a few hours in the country. Within this narrow time span, direct contact with Switzerland and its inhabitants often amounts to little more than a brief interruption of their journey at a petrol station, a road house or an *aire de repos*. The numerous tunnels through the mountains probably merely serve to confirm the "coherent state of mind" of these passers-through that the Swiss motorway and thus Switzerland itself is a transit space that is neither more nor less than the most efficient route from north to south and vice versa, with the bonus of superb views.] [Faure's photographs show that not just the motorway but the entire motorway landscape can be conceived as a "single comprehensible place" that is unrelated to the rest of the country. For a start, the crash and noise barriers which line the road and render the motorway landscape impenetrable to the surrounding countryside, ensure that the road and its verges are autonomous, a cordoned-off corridor through town and country.] [The vegetation of the motorway landscape accentuates this contrast. The flora is selected according to its resistance to air pollution. It is preferably low-maintenance and does not grow or spread so rampantly as to cause any kind of hindrance or danger to traffic. These pragmatic considerations are combined with a landscape architecture that is in many respects about as natural as the average golf course. Another thing the designed motorway landscape has in common with the golf course is its repetitive, interchangeable and often completely decontextualized manifestation, which has no relationship with the surrounding landscape save for the fact that it occupies part of its space.] [The ponds and mini-marshes beside some stretches of motorway are not natural either. They are the result of negotiation with anti-motorway protestors, a form of enforced compensation for the nature lost through the construction and widening of the roads. In their isolation they proclaim all the more clearly their status as atonement-nature, a penance for the environmentally polluting sins of road builders.] [All of which is no reason to be condescending about the motorway landscape. For it is also clear that this landscape is designed with care and skill. That the result of these efforts is unspectacular is not a shortcoming but deliberately contrived in the interests of road safety. This is why on-ramps, for example, are never lined with trees because they could obstruct the view. This is less of a problem with off-ramps and so these are sometimes flanked by trees.] [The photographs Faure has taken of this remarkable landscape undeniably afford aesthetic pleasure. Even if what can be seen is by no means always beautiful and by no means always meant to be beautiful. Nor does it seem to be Faure's intention to make the viewer aware of the beauty of the motorway landscape; he simply wants to show it as it is. But just as a good journalist is able to express bad news in fine words, so Faure is only able to show the banality of this motorway landscape in a beautiful manner.] [If in so doing he opens the viewer's eyes, it is not so much to the beauty as to an awareness of the omnipresence of the kind of landscape he has photographed. Far from being confined to motorways, it also appears along railway lines, in particular the TGV, on and around industrial and business estates, in office districts, around shopping centres and supermarkets, around airports and so on and so forth. In all these sorts of environments onc finds a similarly efficient and pragmatic greenery. The only thing to be said for it is that it absorbs carbon dioxide and produces oxygen, that it fills a void and consoles all those who believe that greenery is more natural and thus more human than asphalt and concrete. In that respect, these photographs of Faure's are charged with a social and political message. Without being explicitly judgmental, he nevertheless manages to show quite clearly the consequences—unintended and probably also unforeseen—of the way of life that the Swiss in particular, but also West Europeans and other inhabitants of wealthy parts of the world, are able to afford at the beginning of the 21st century.

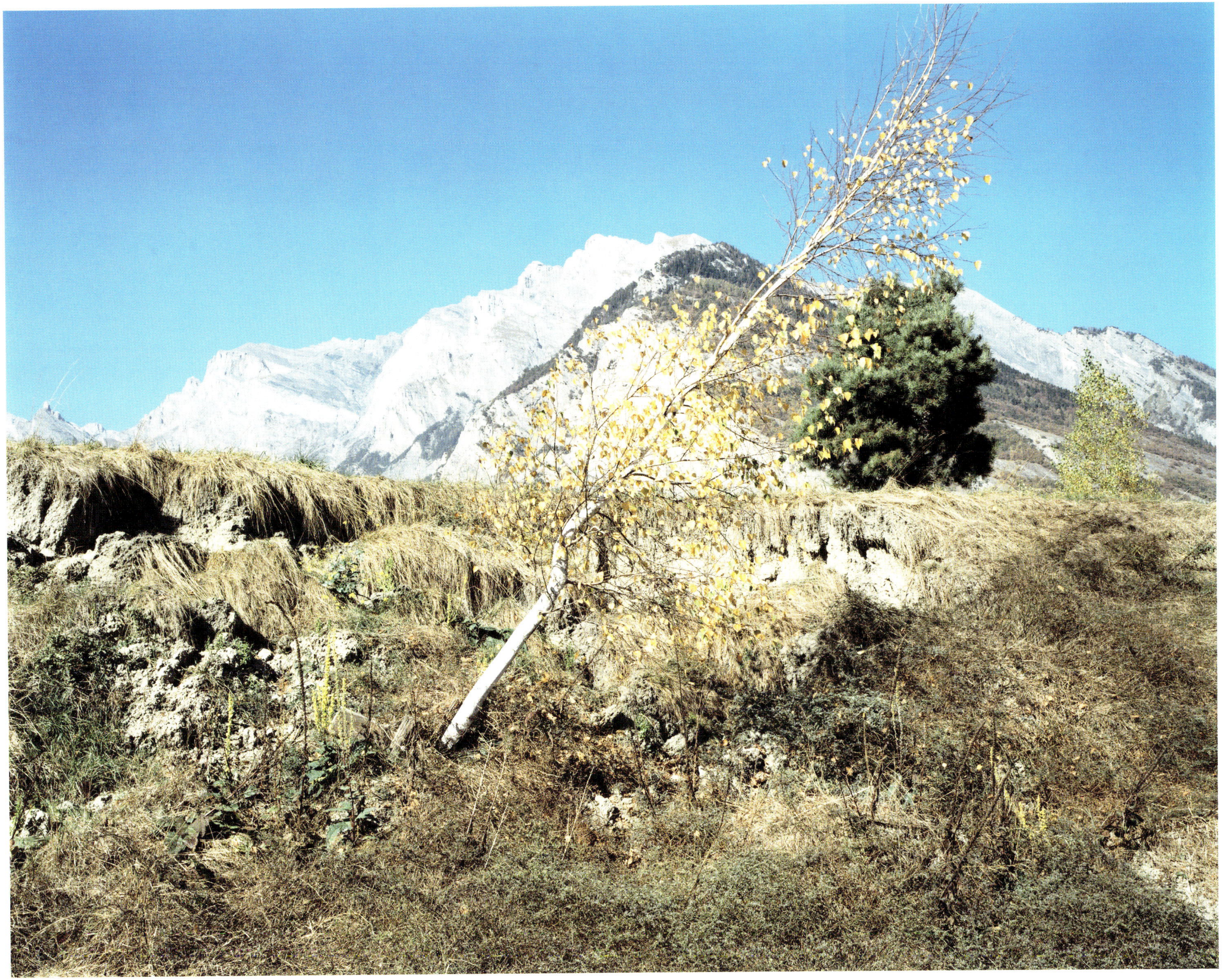

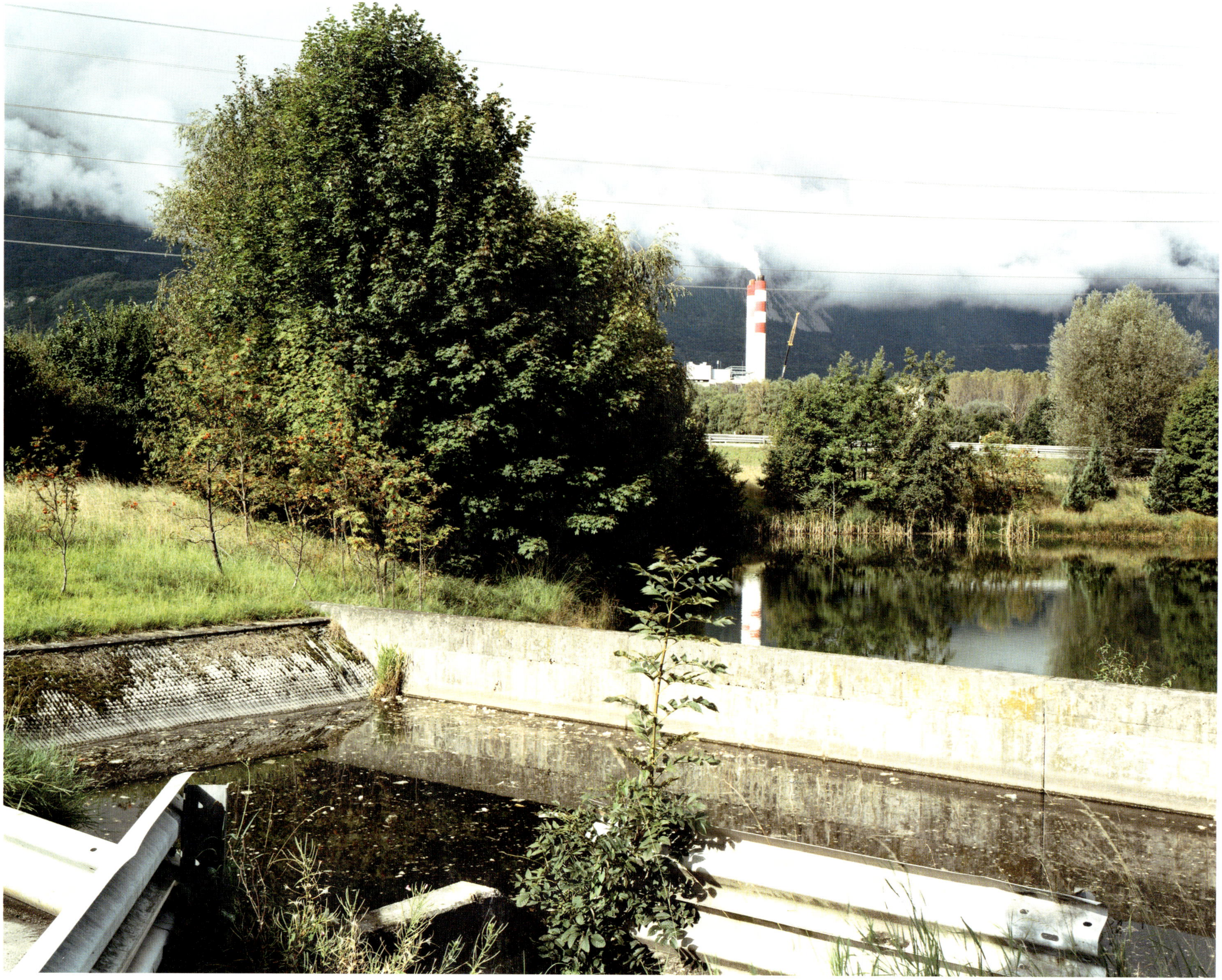

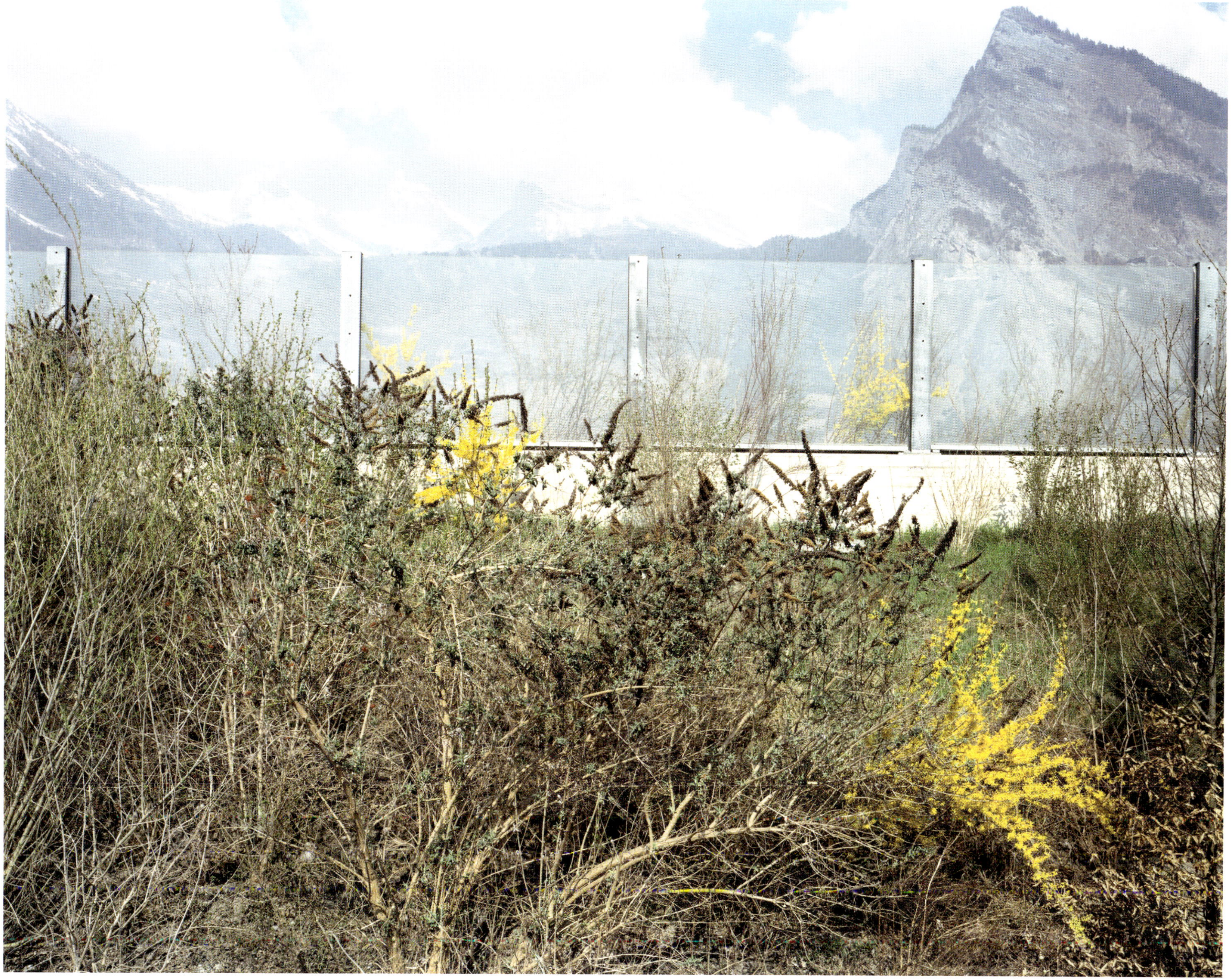

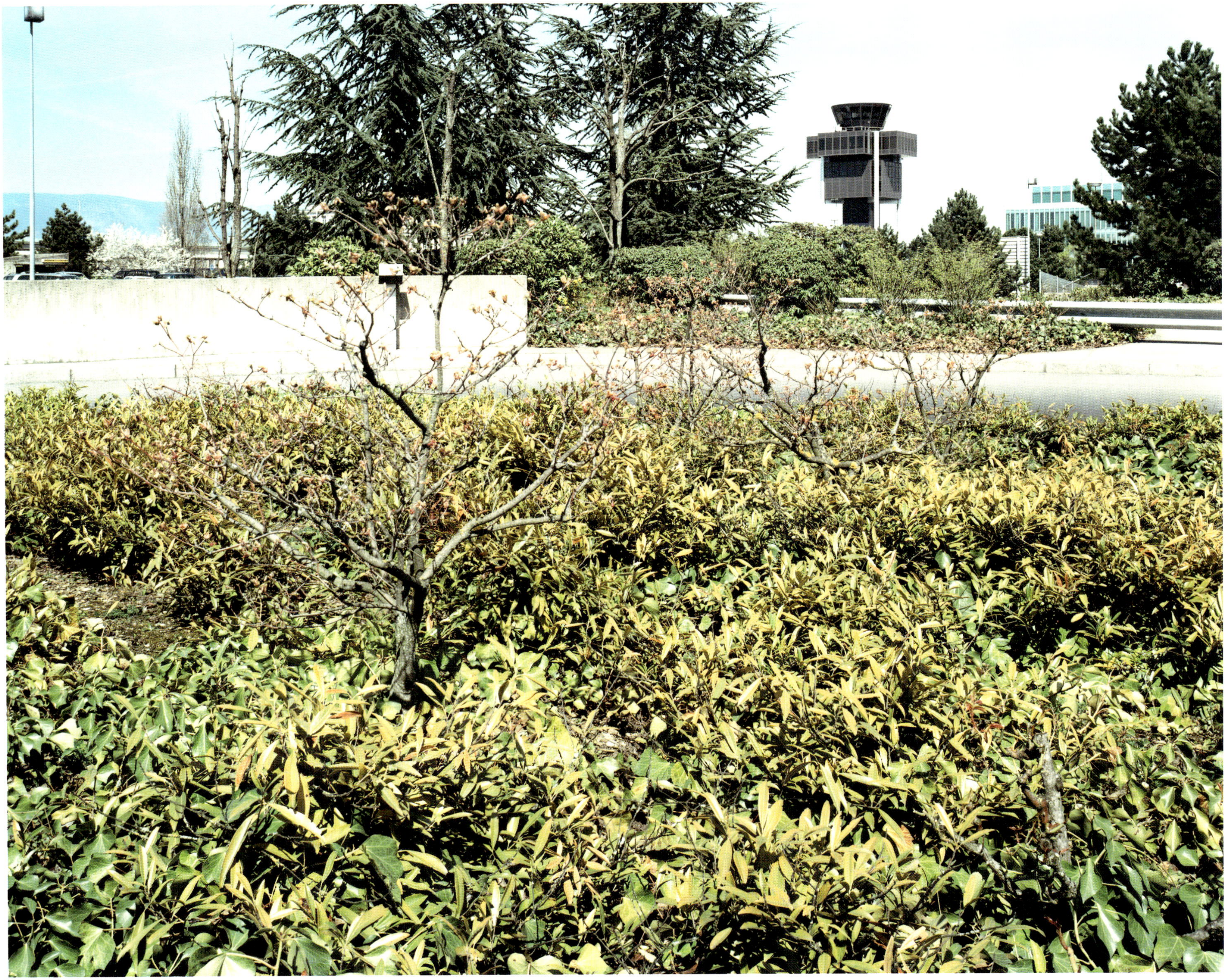

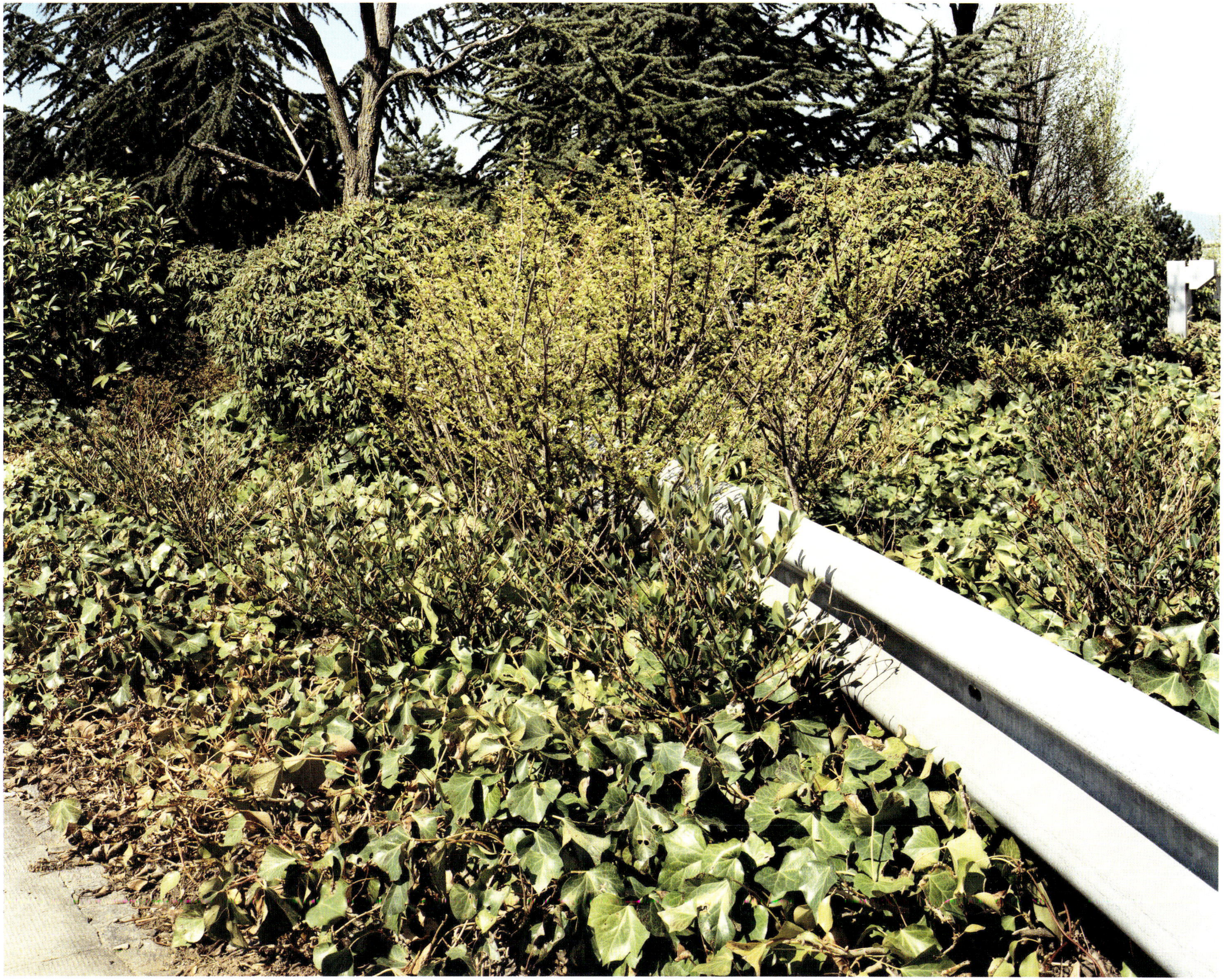

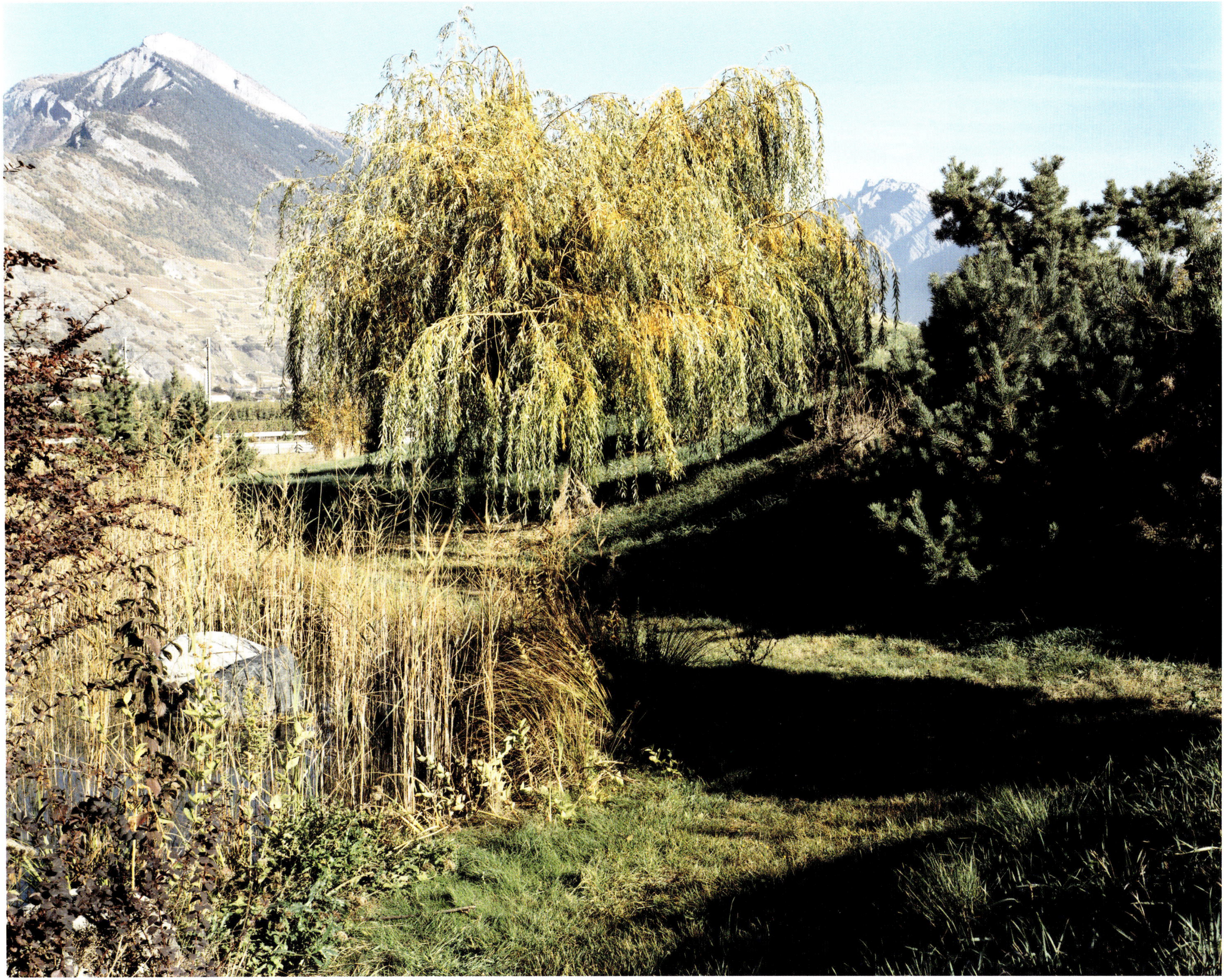

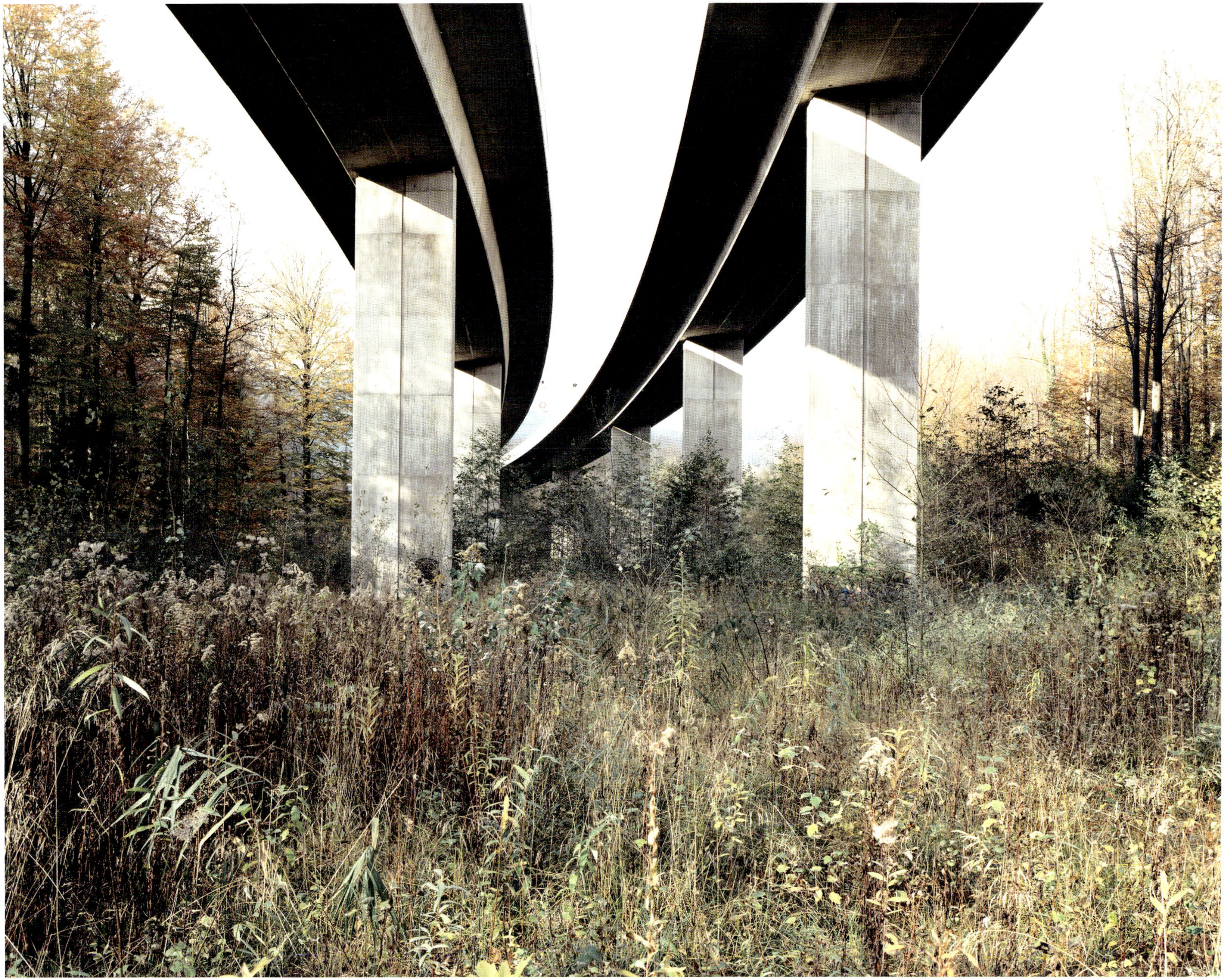

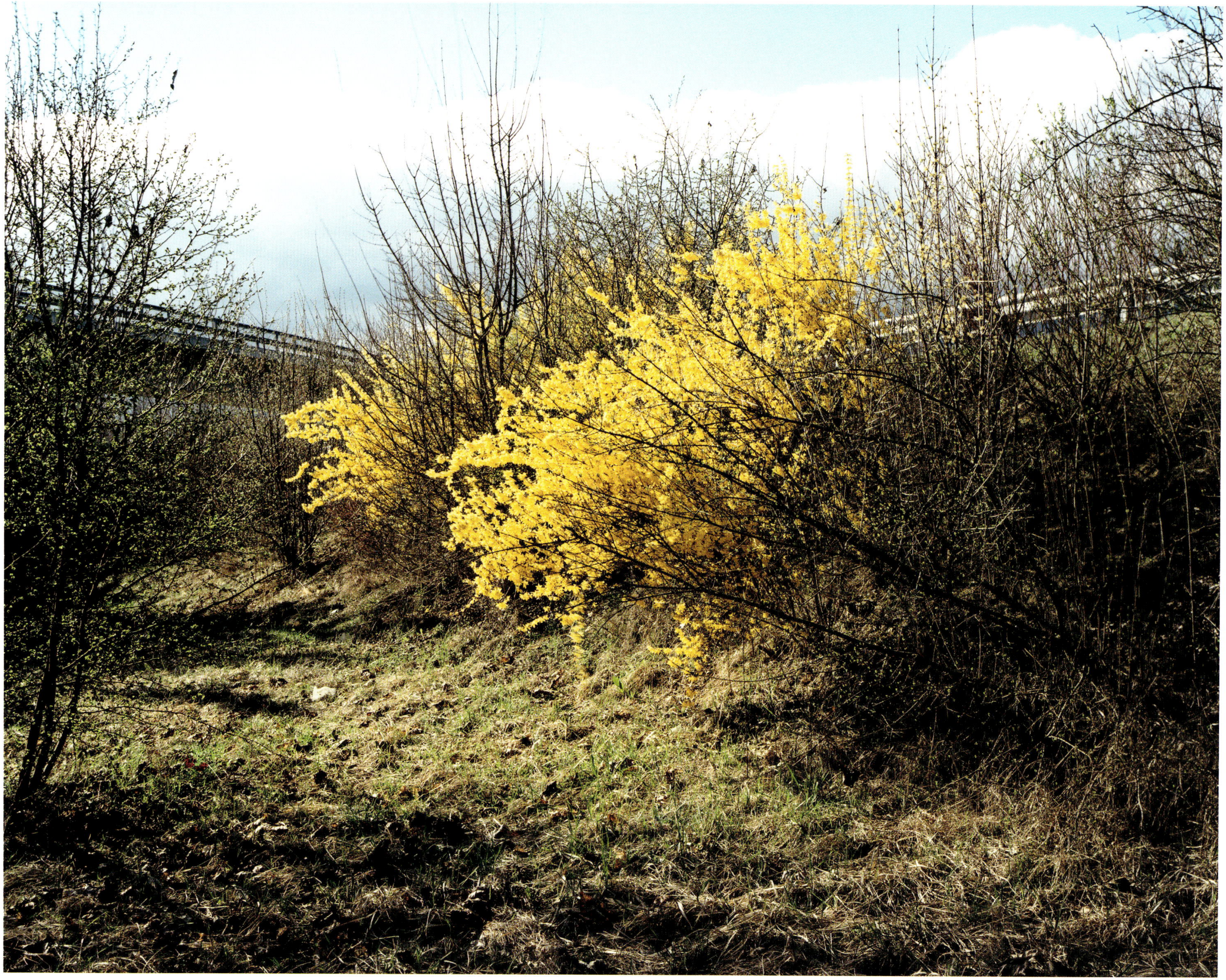

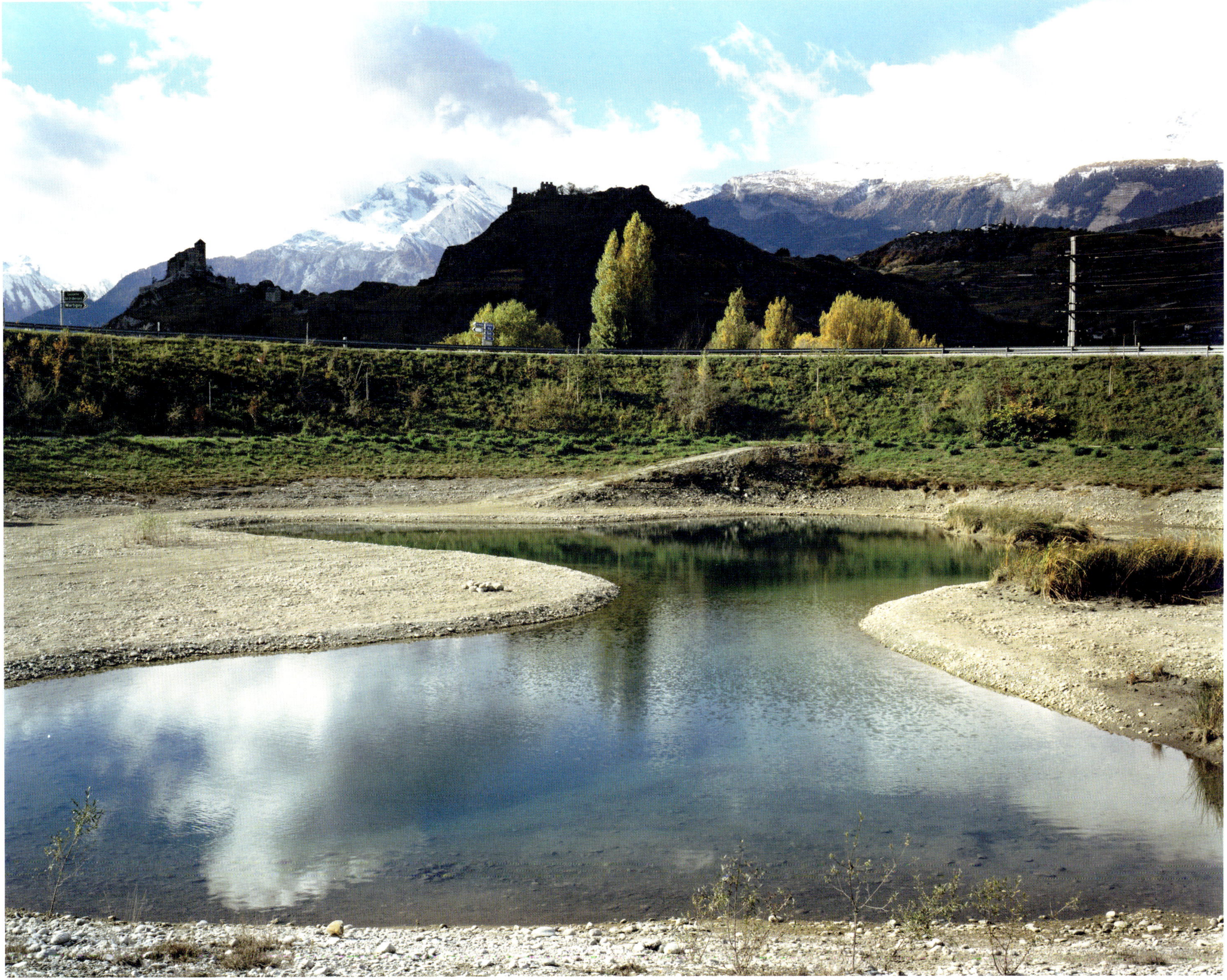
Martigny

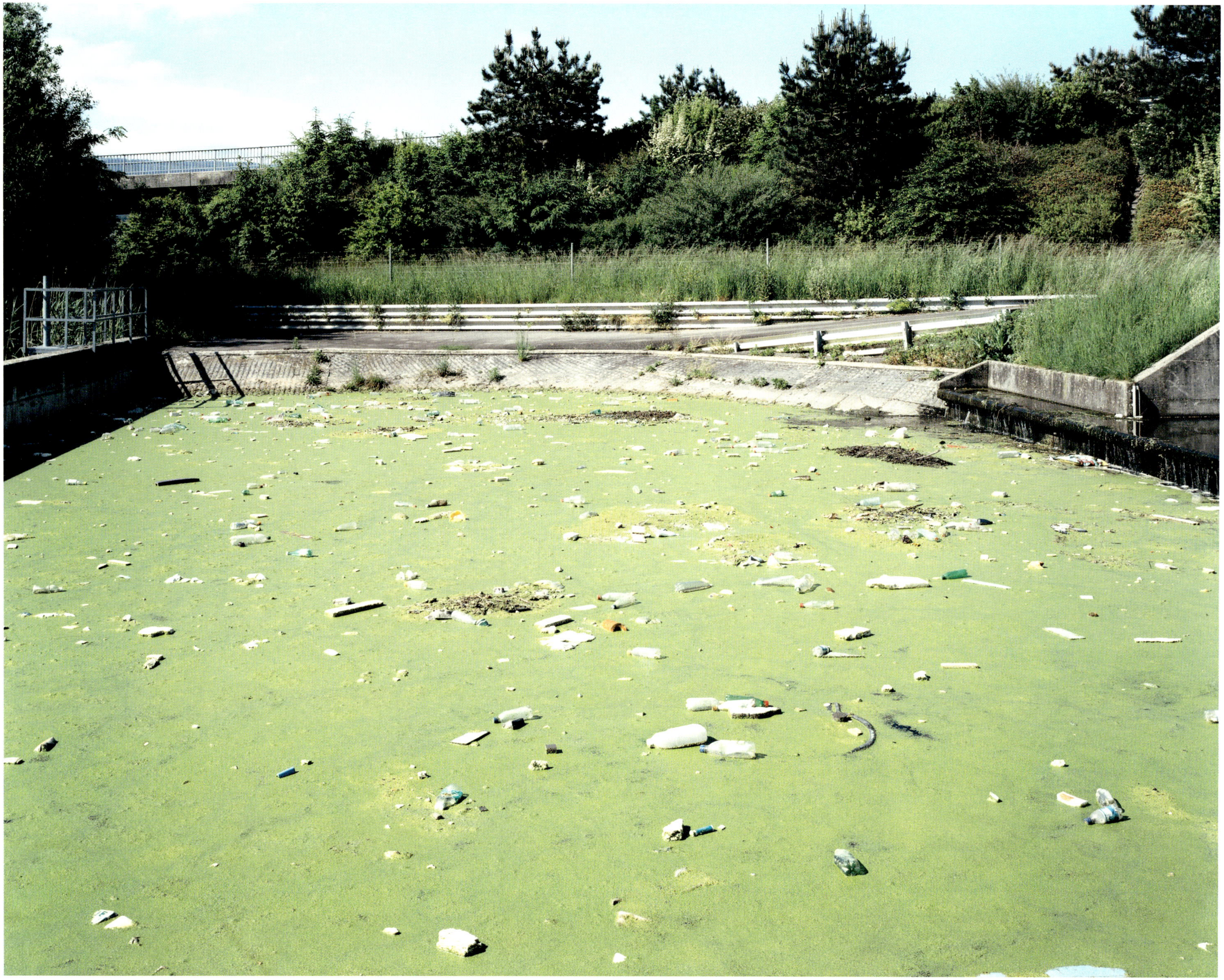

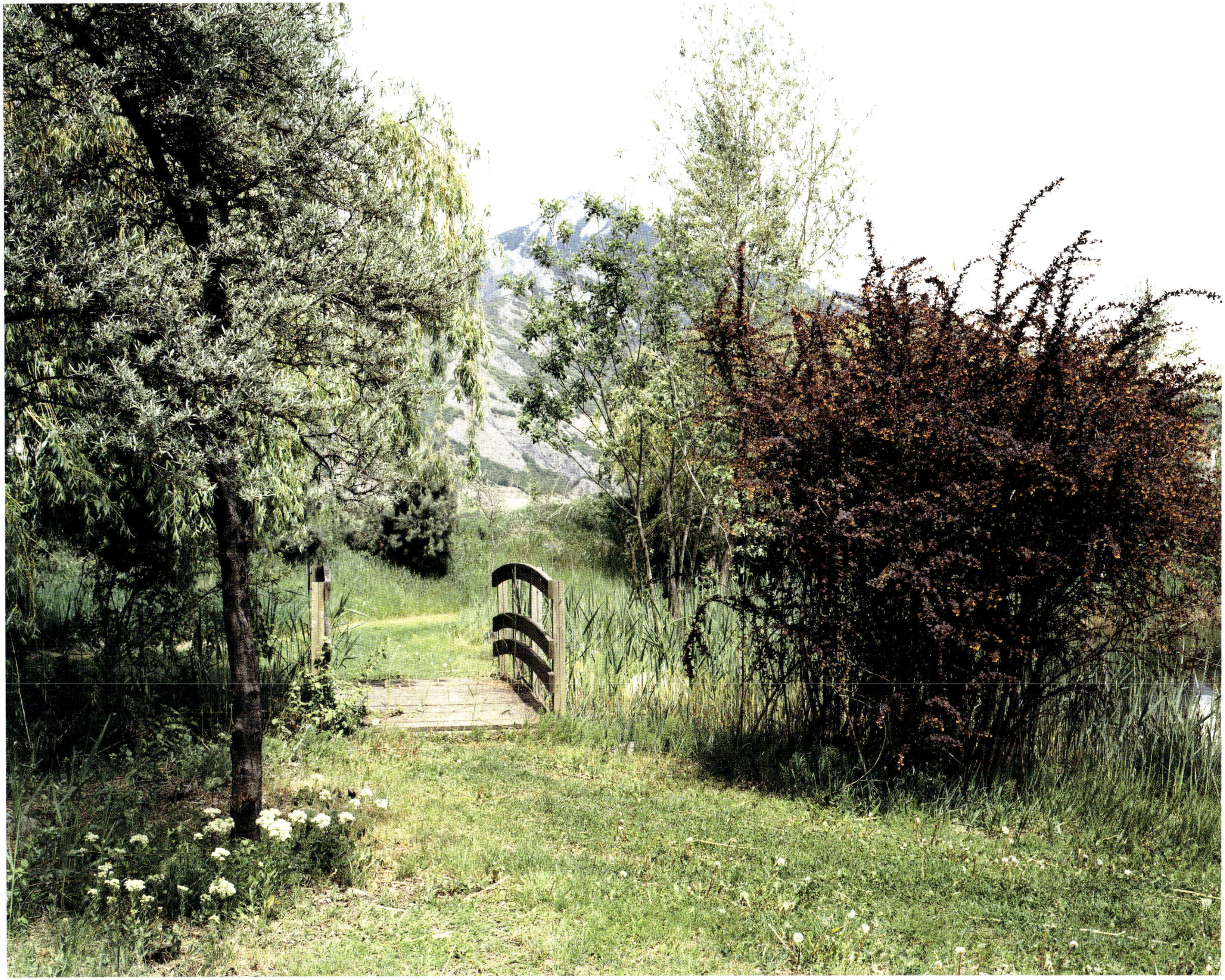

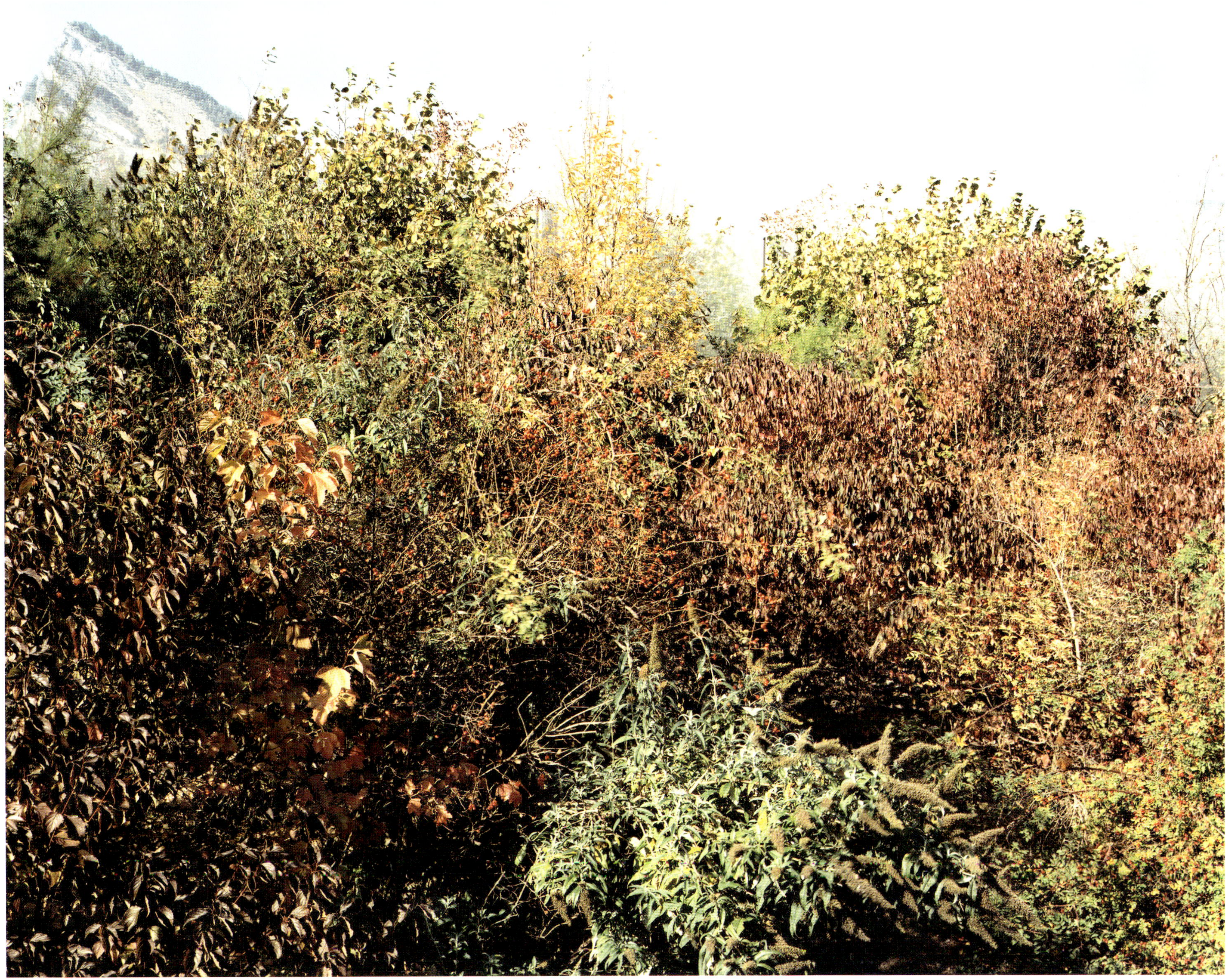

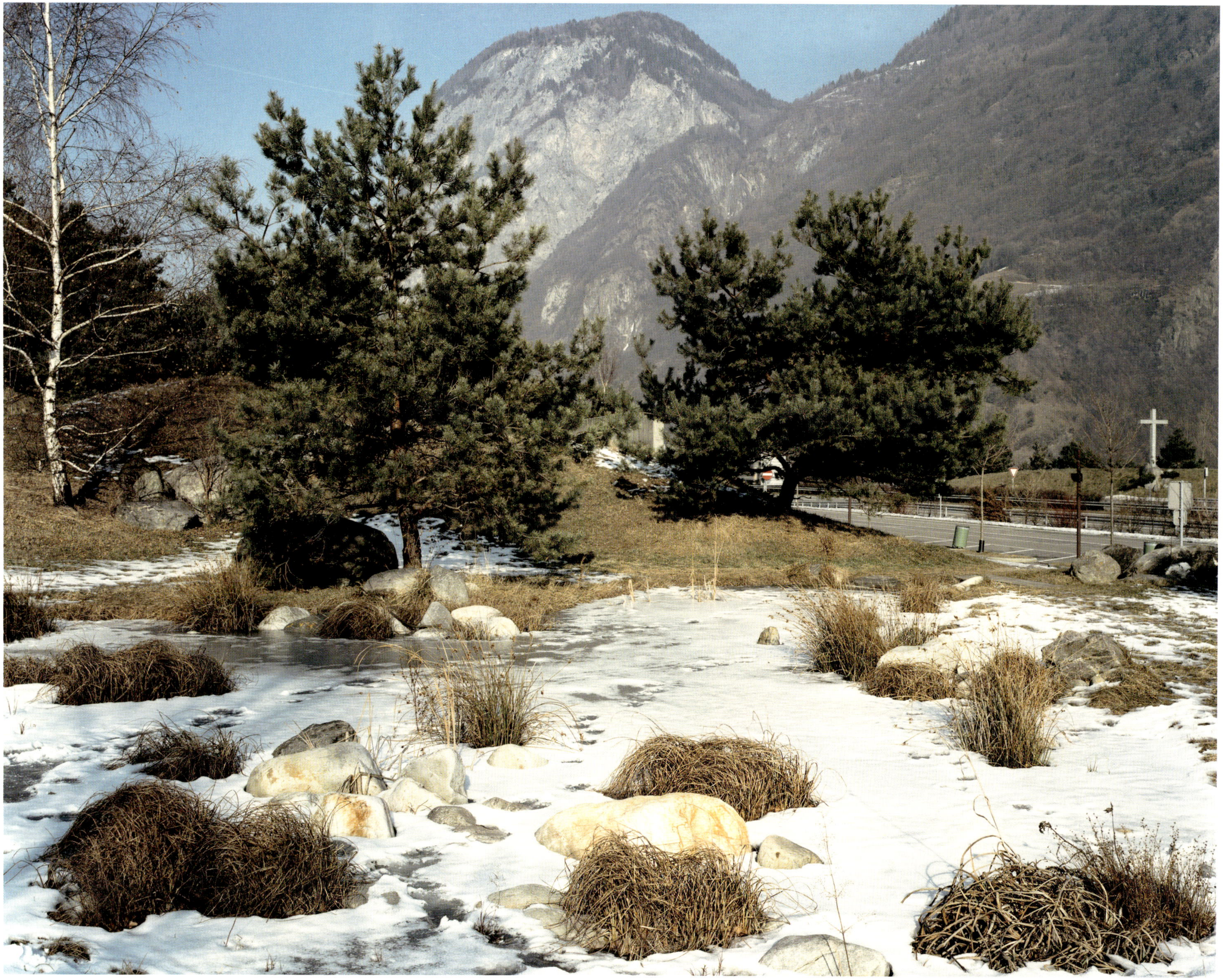

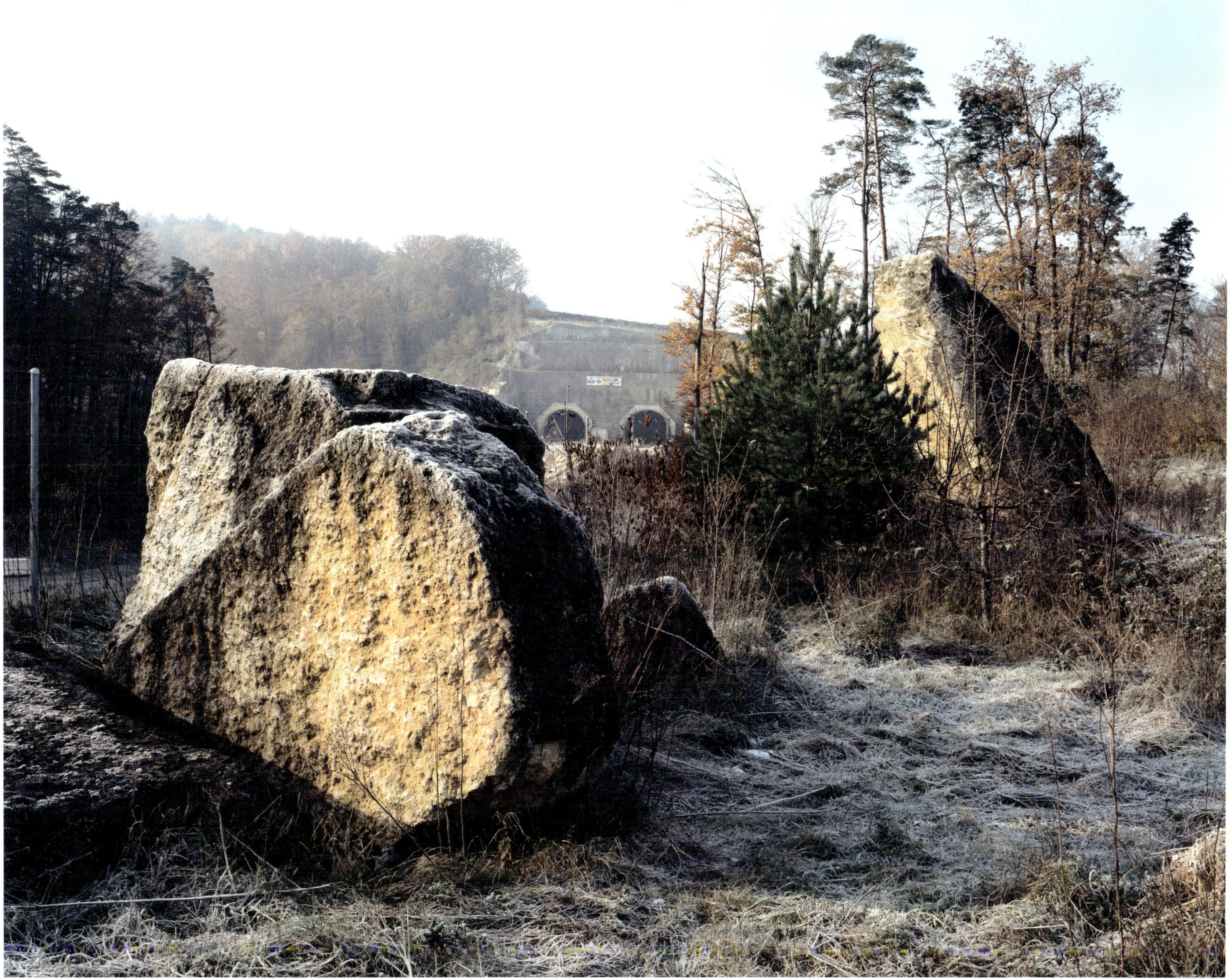

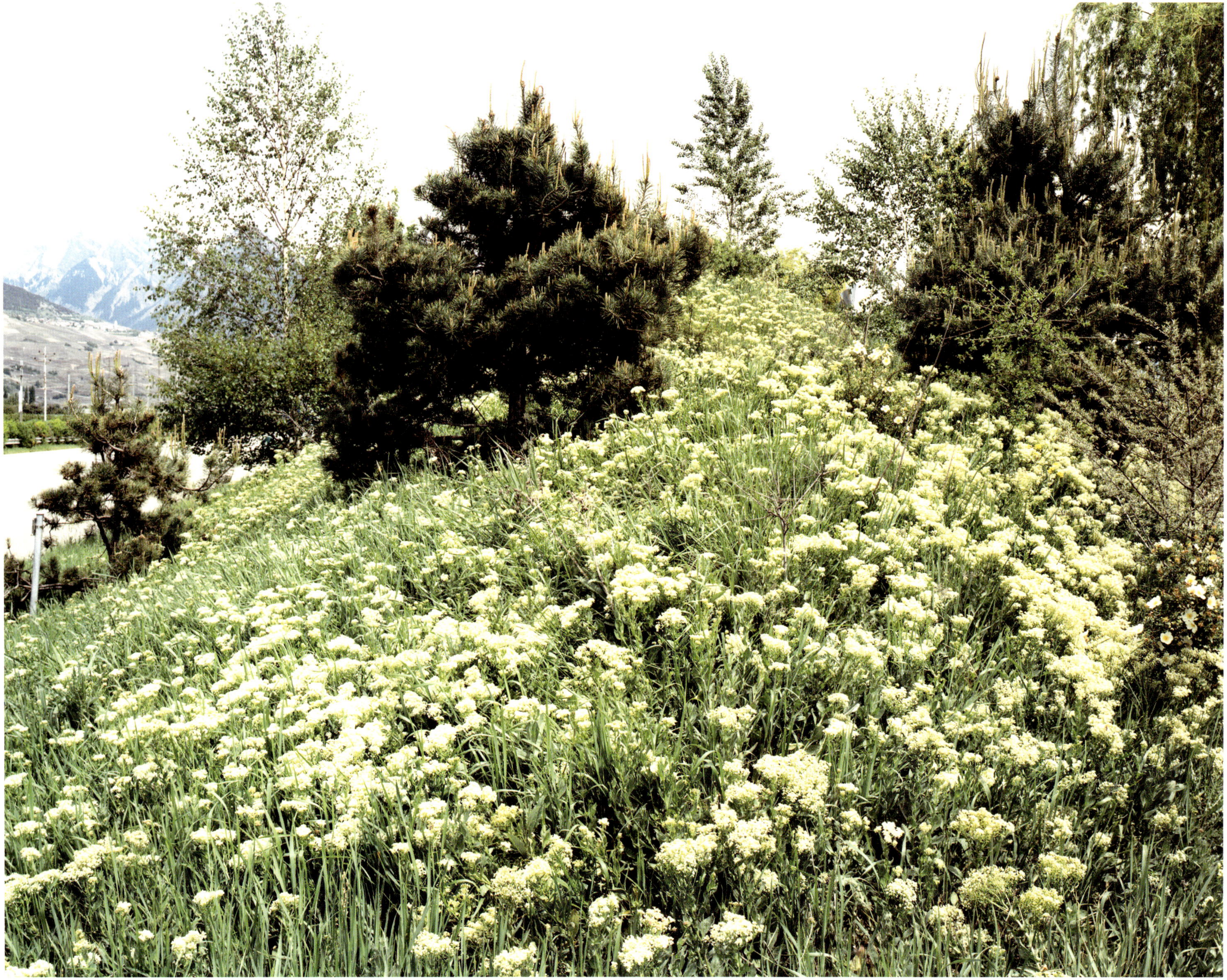

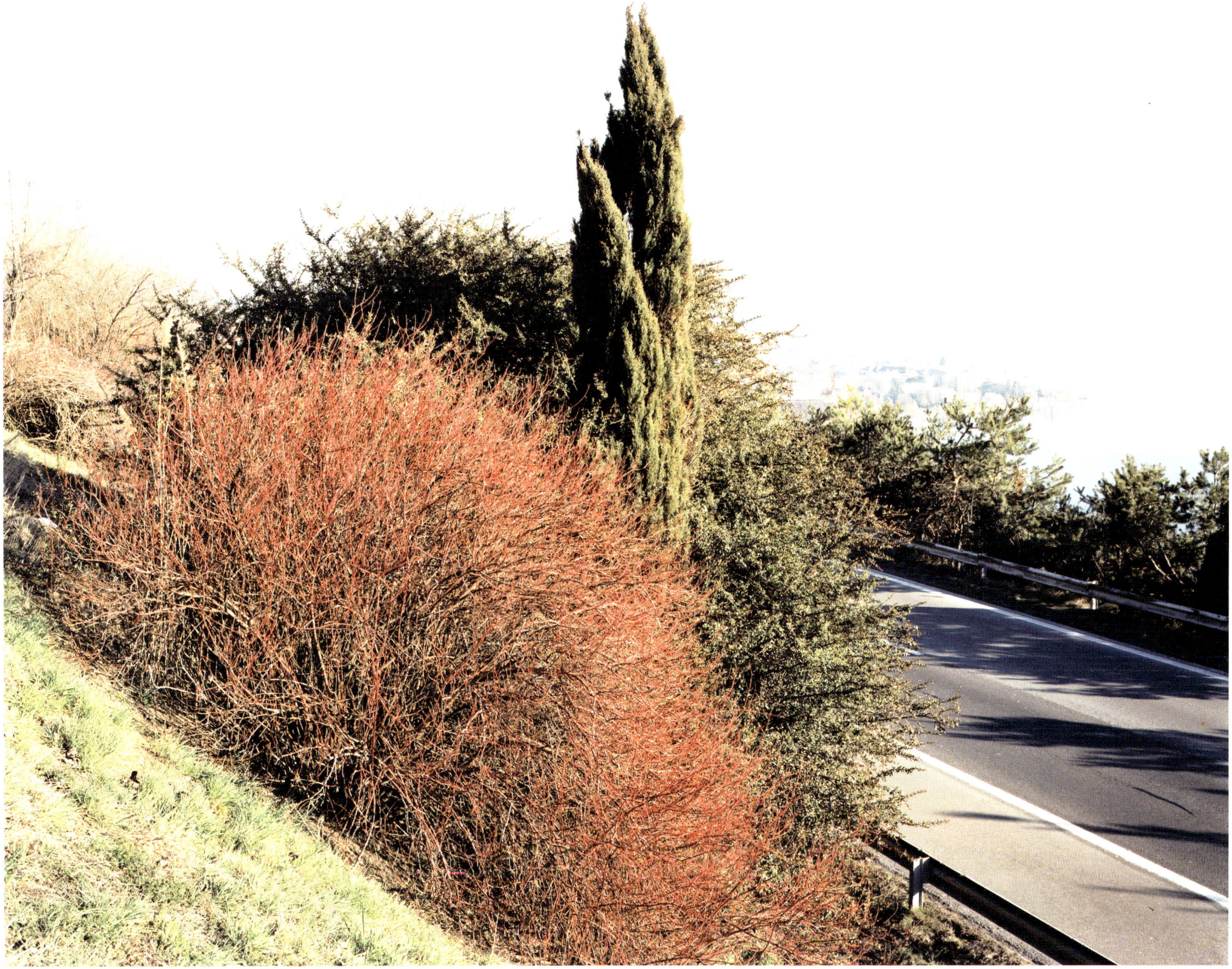

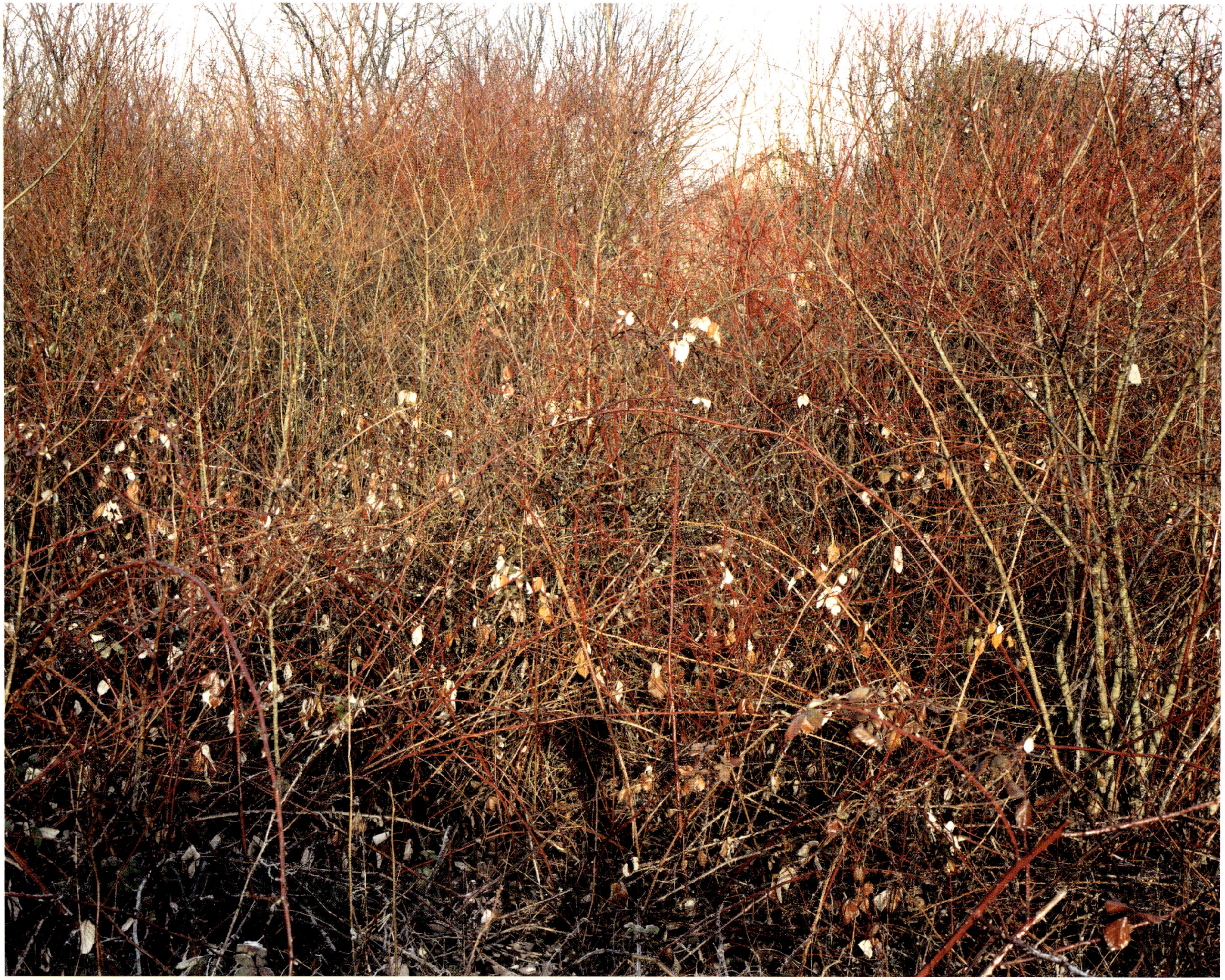

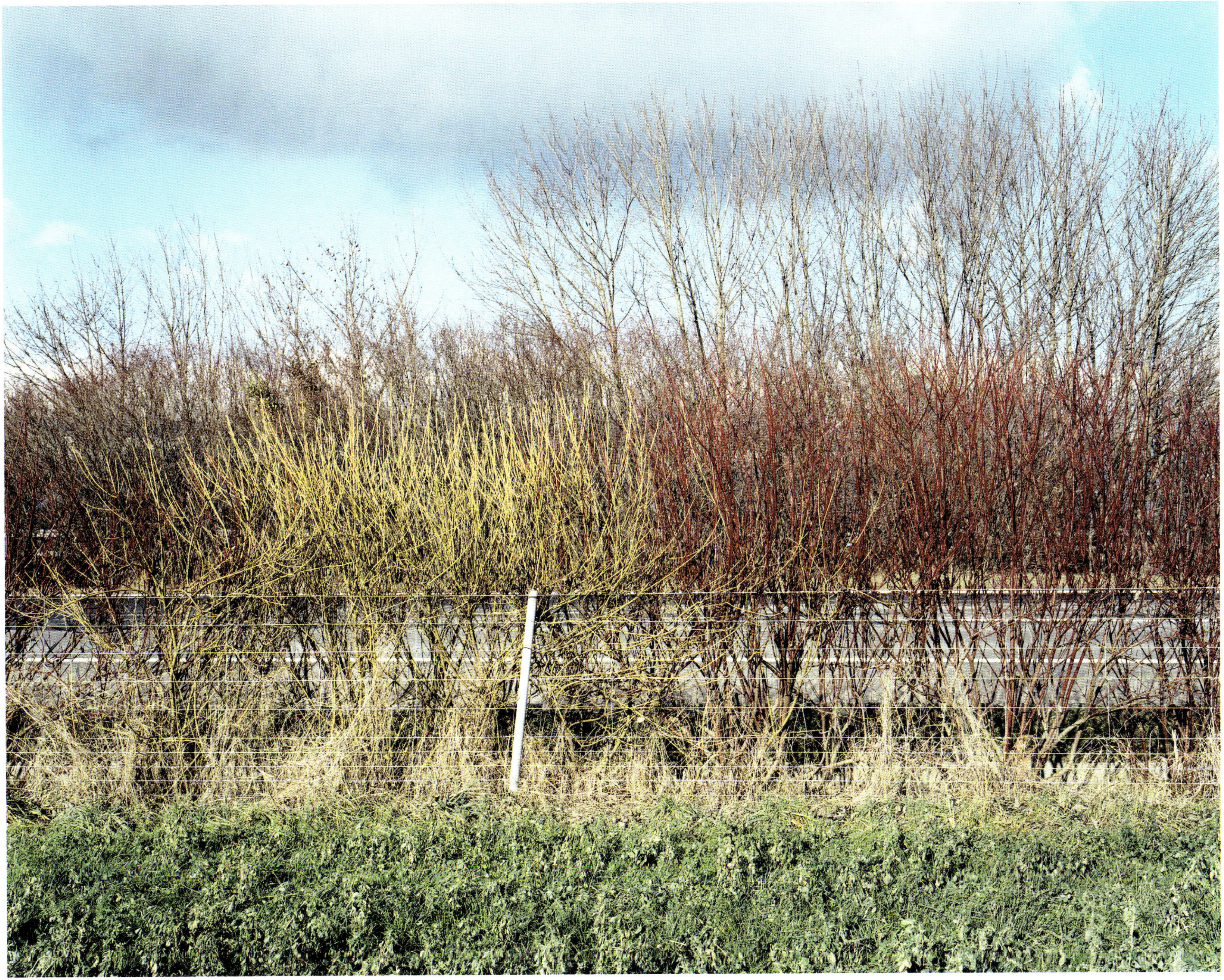

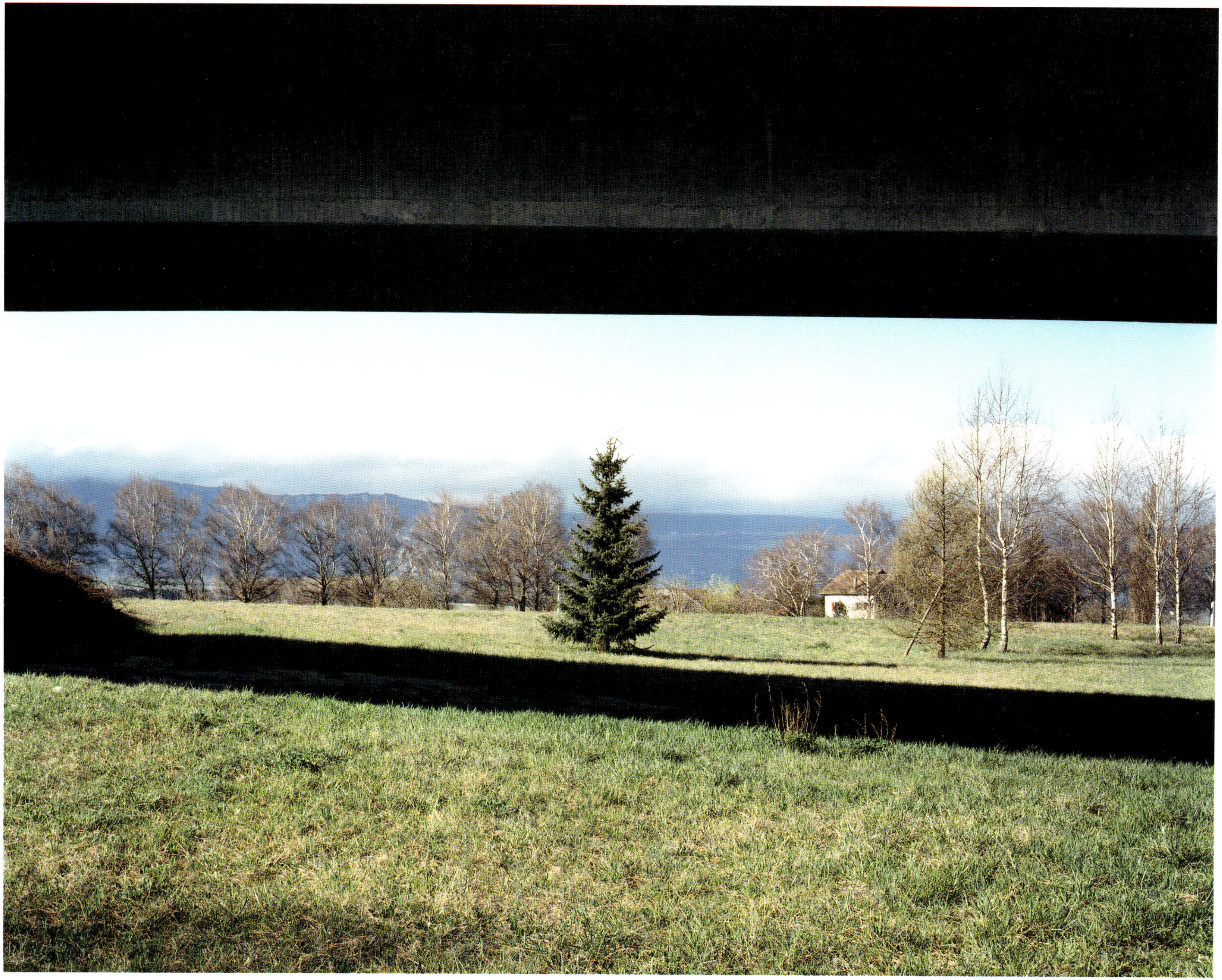

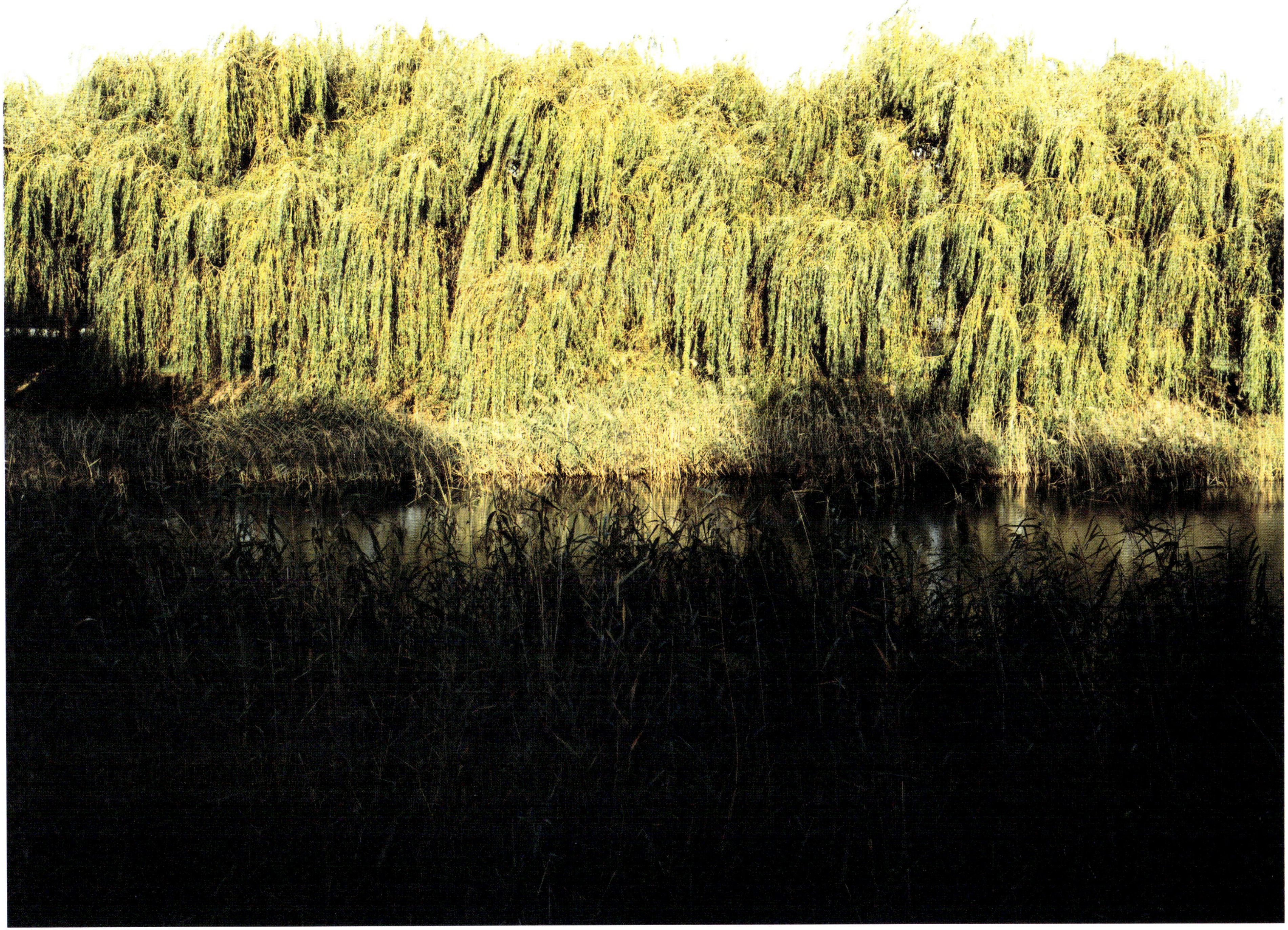

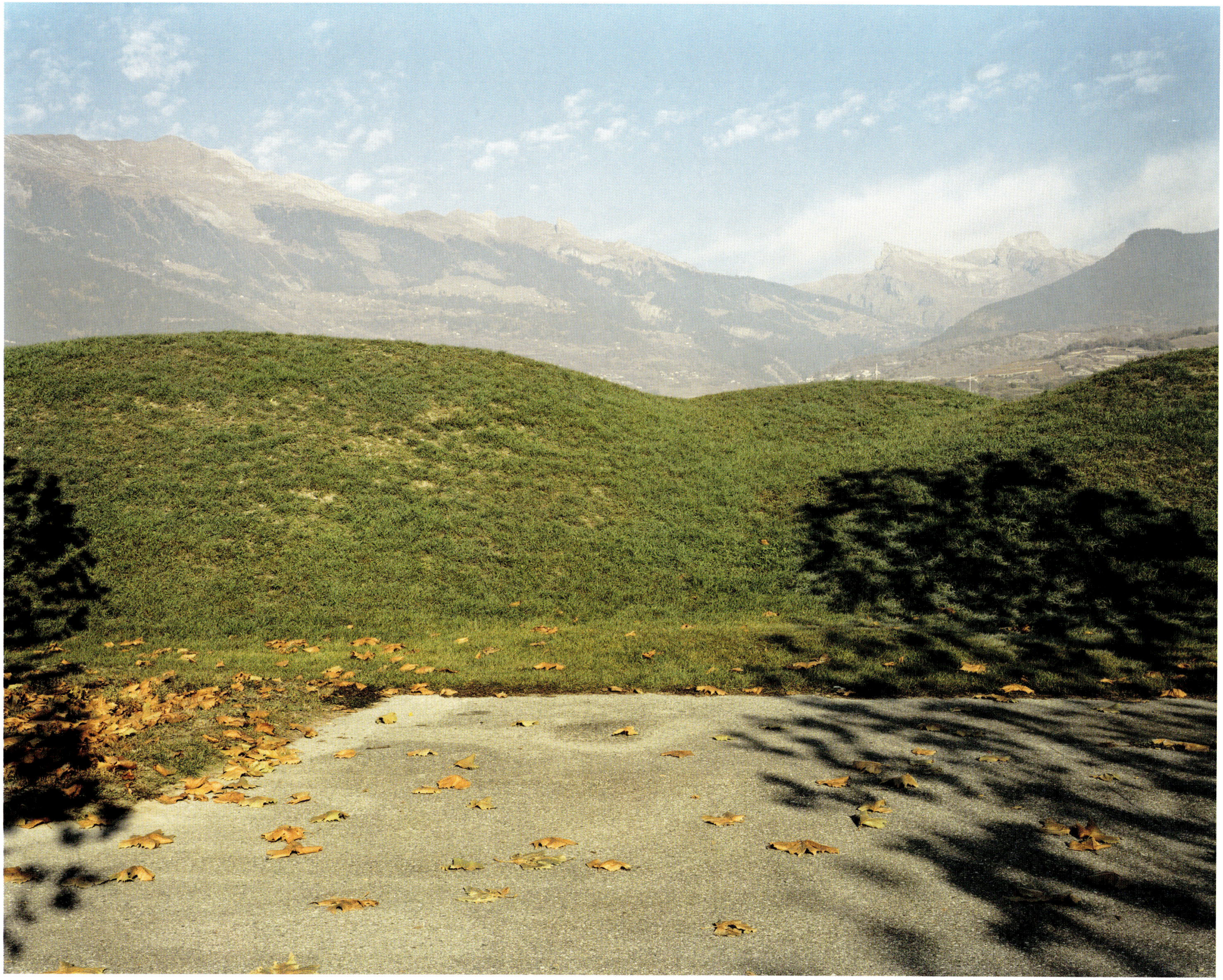

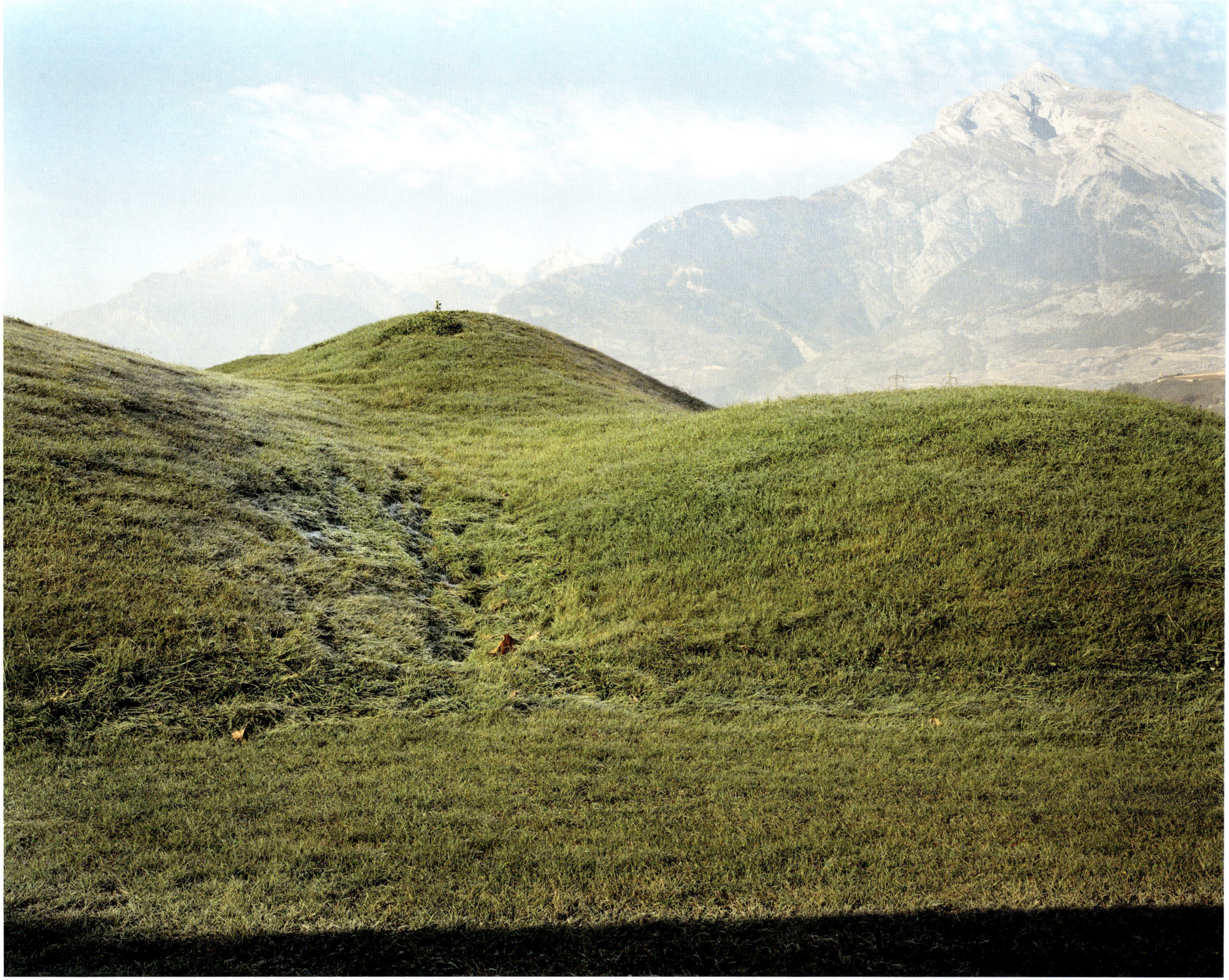

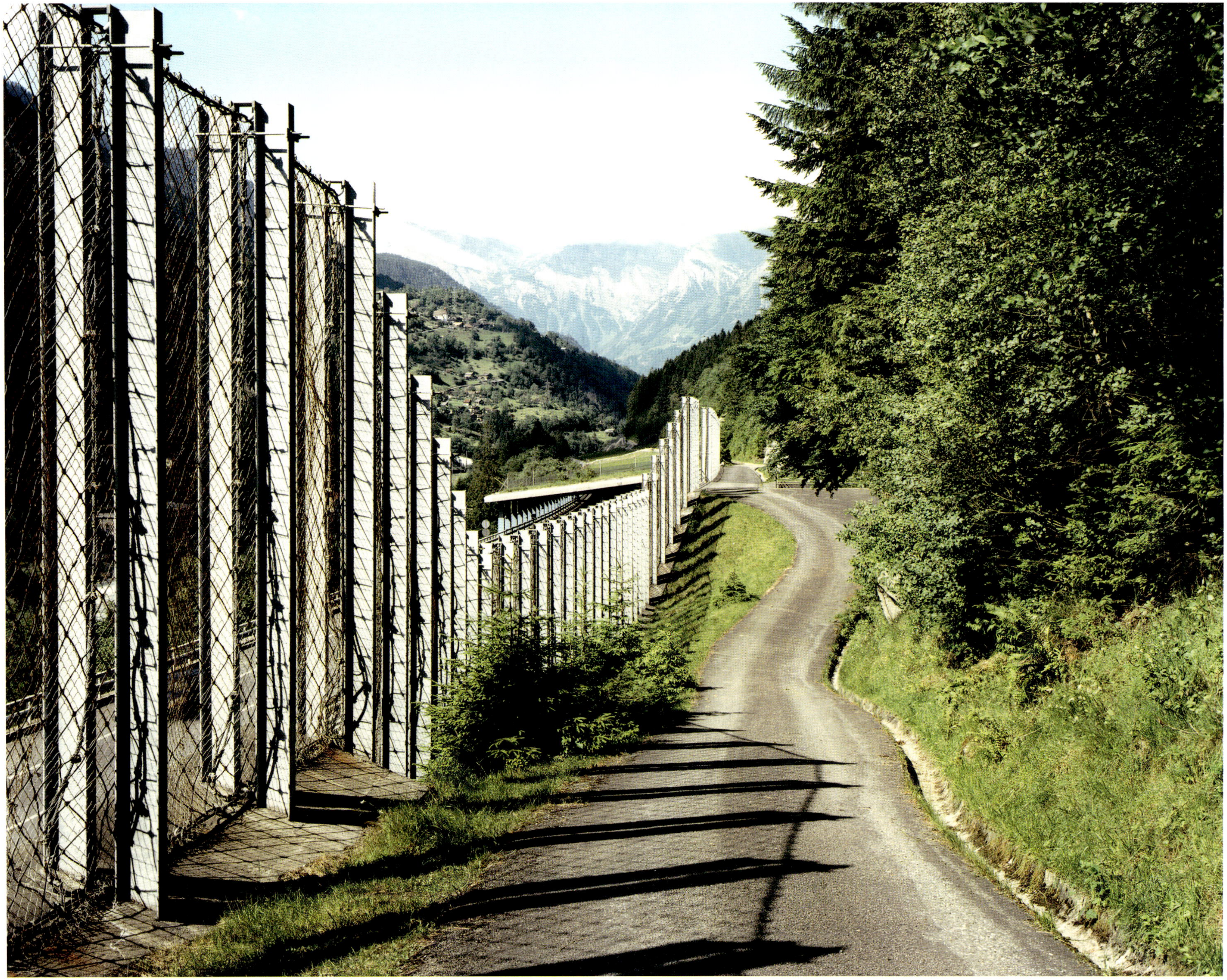

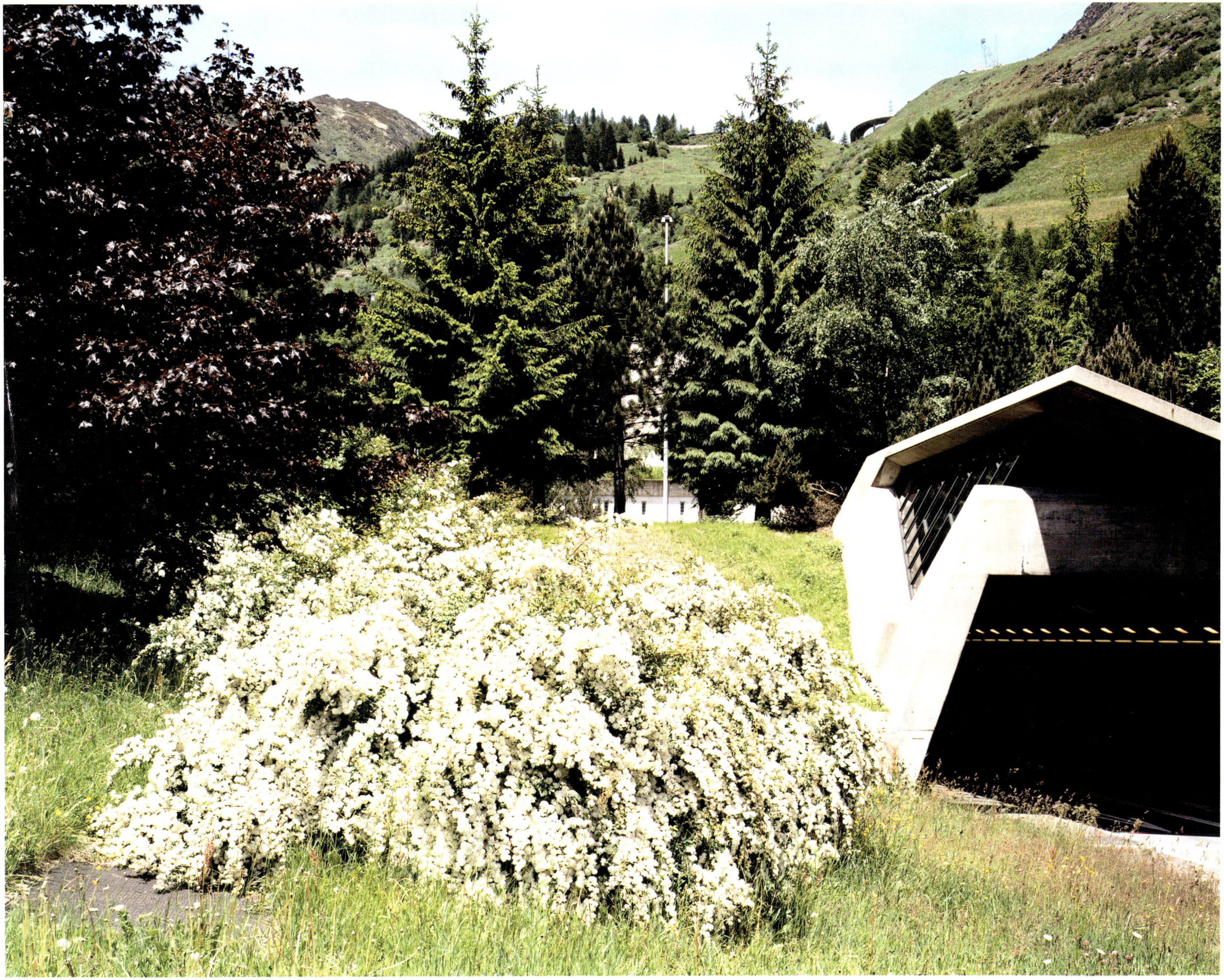

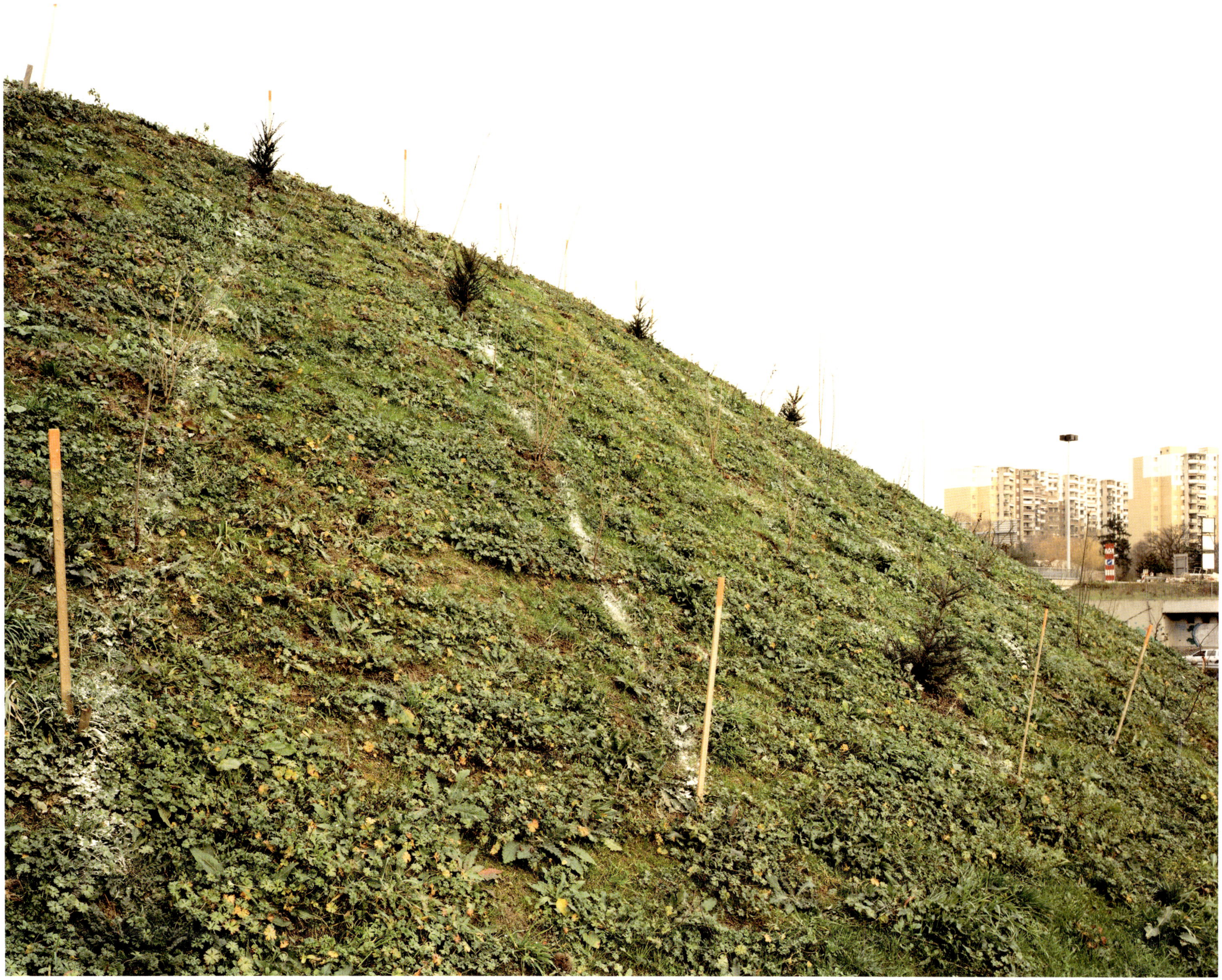

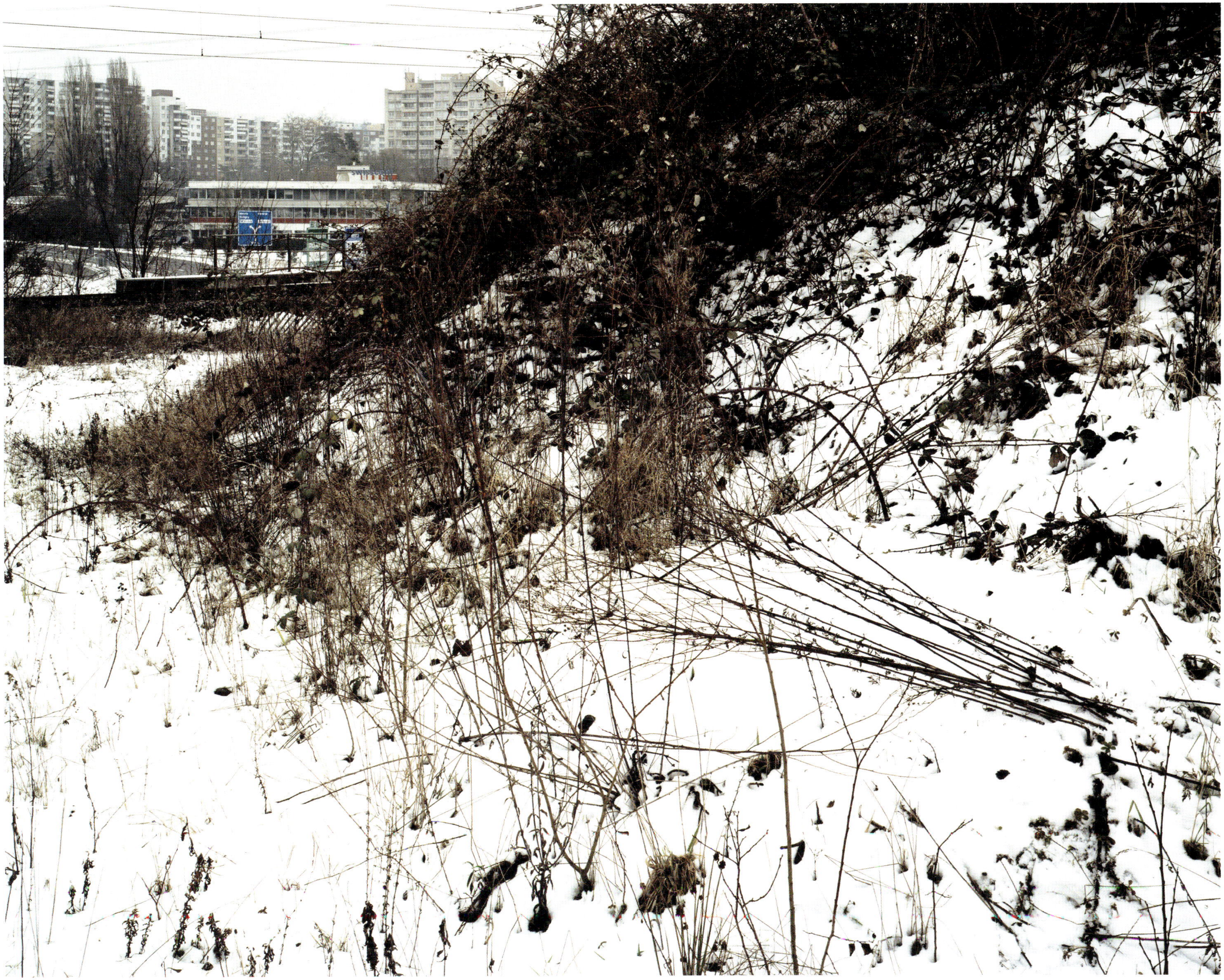

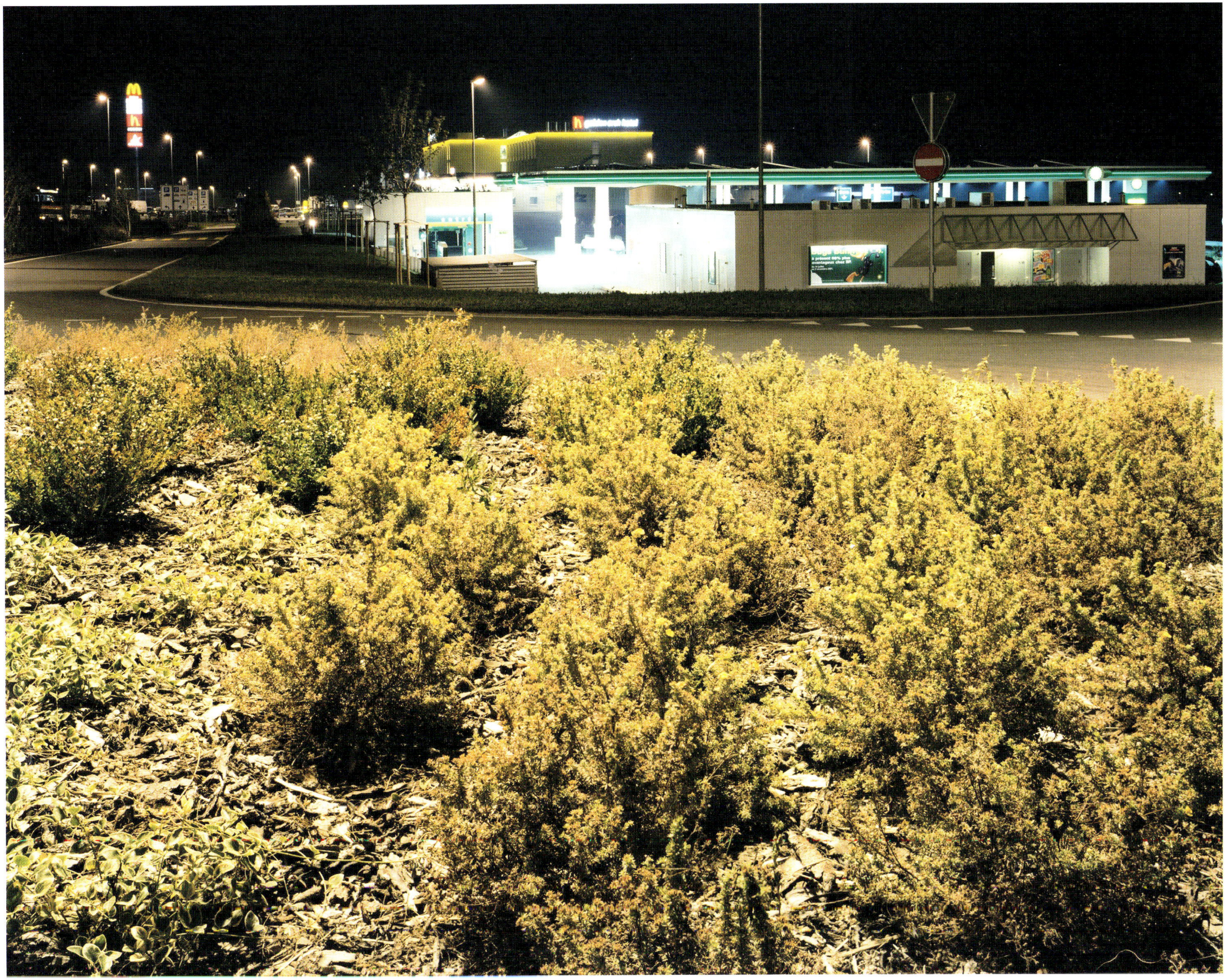

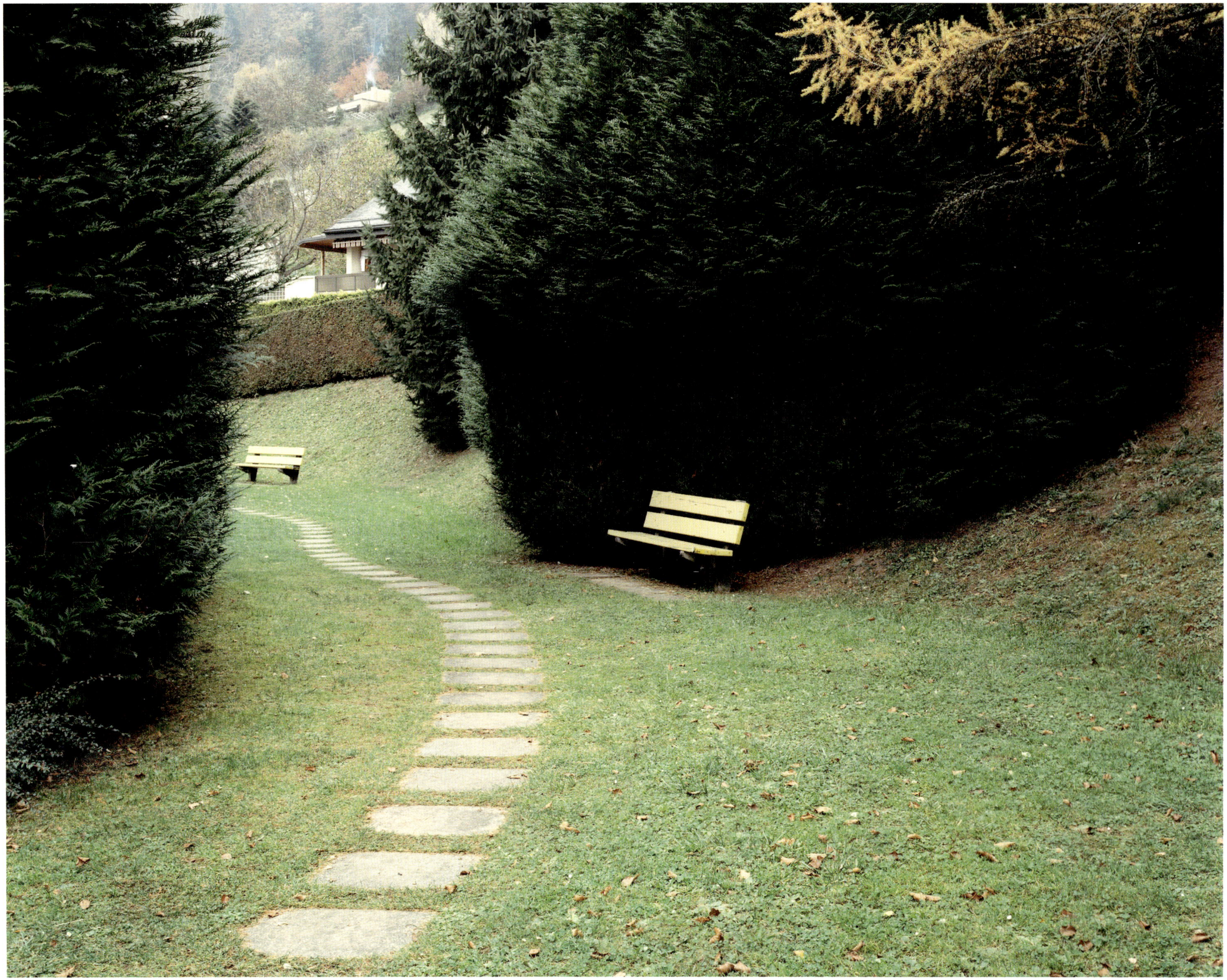

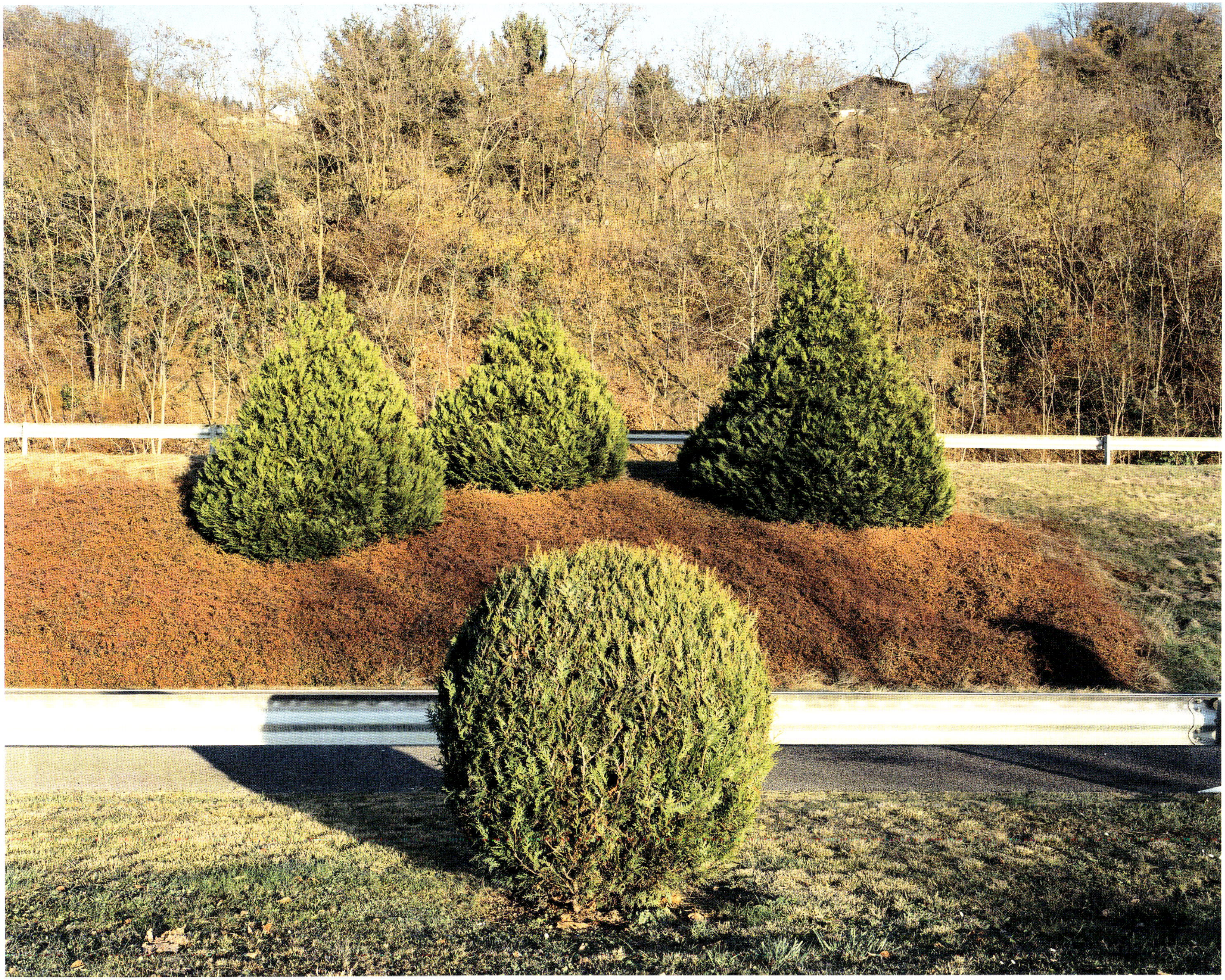

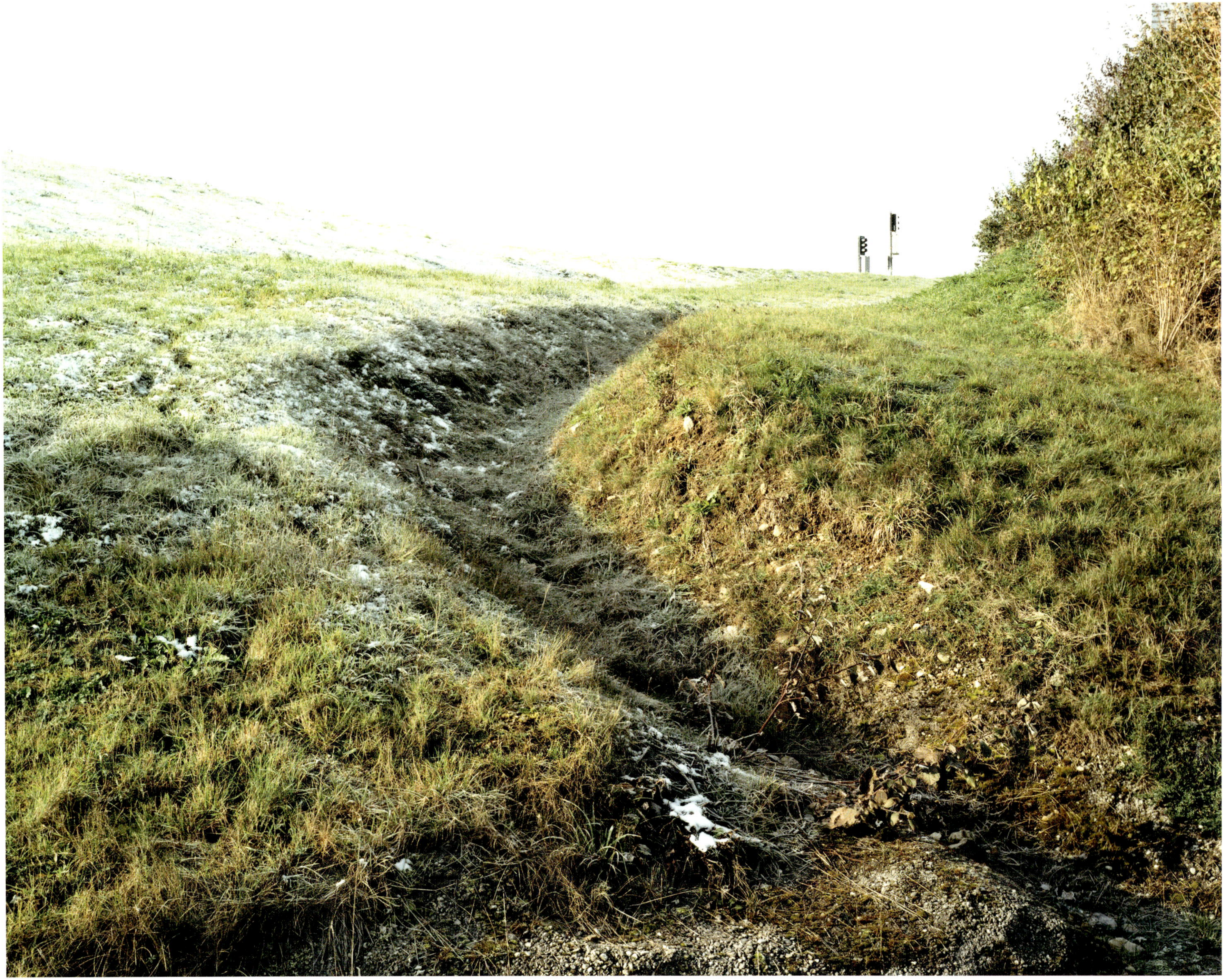

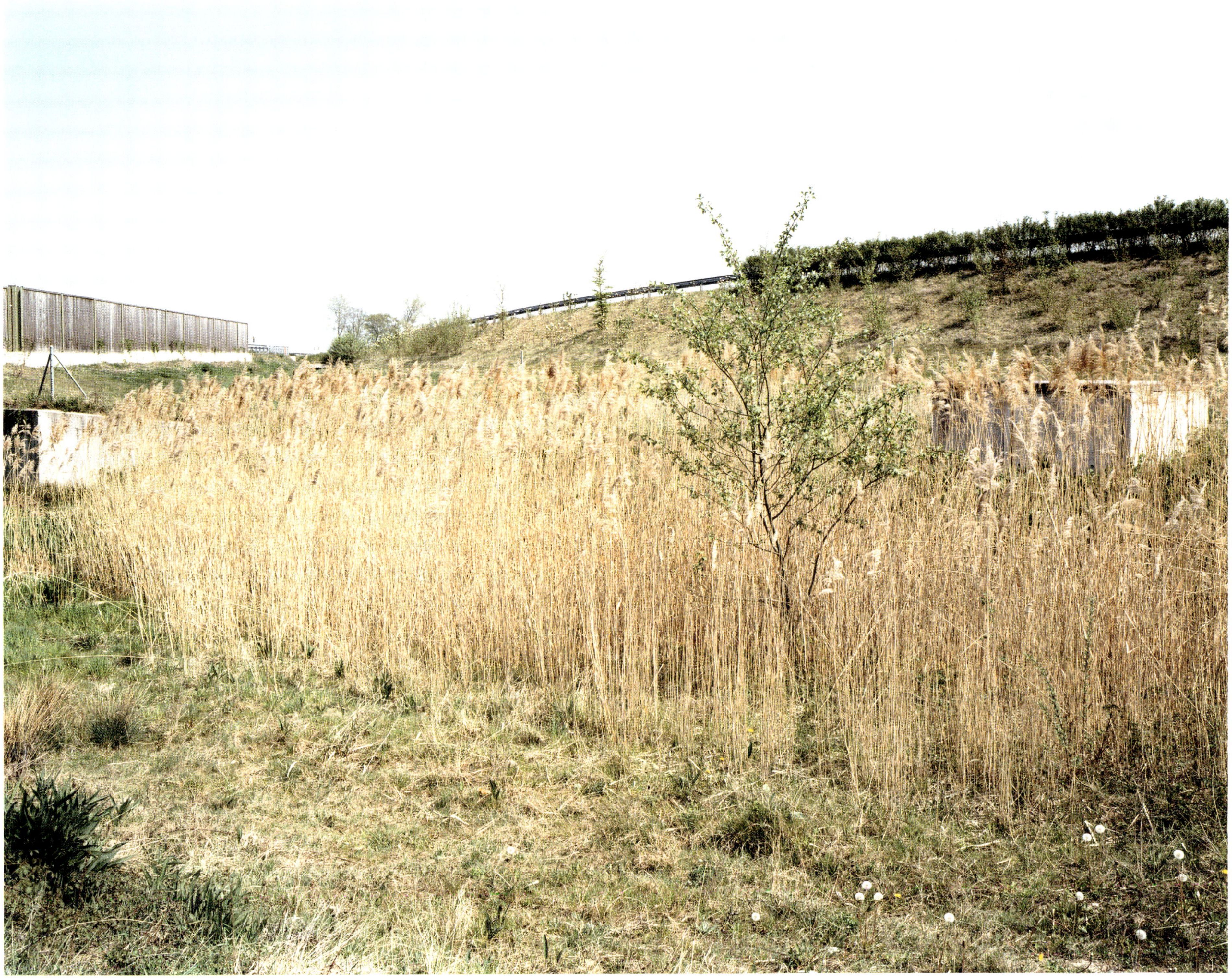

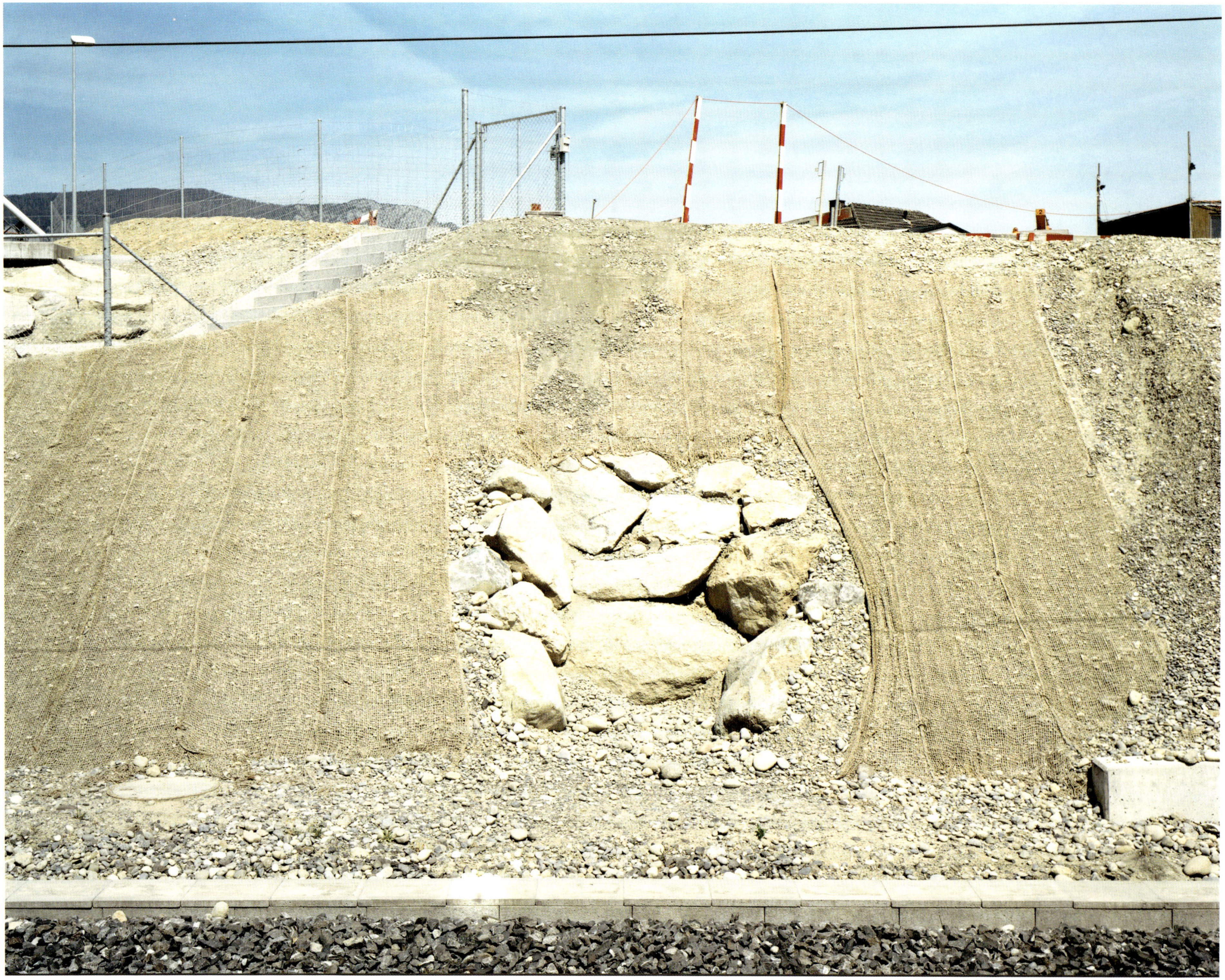

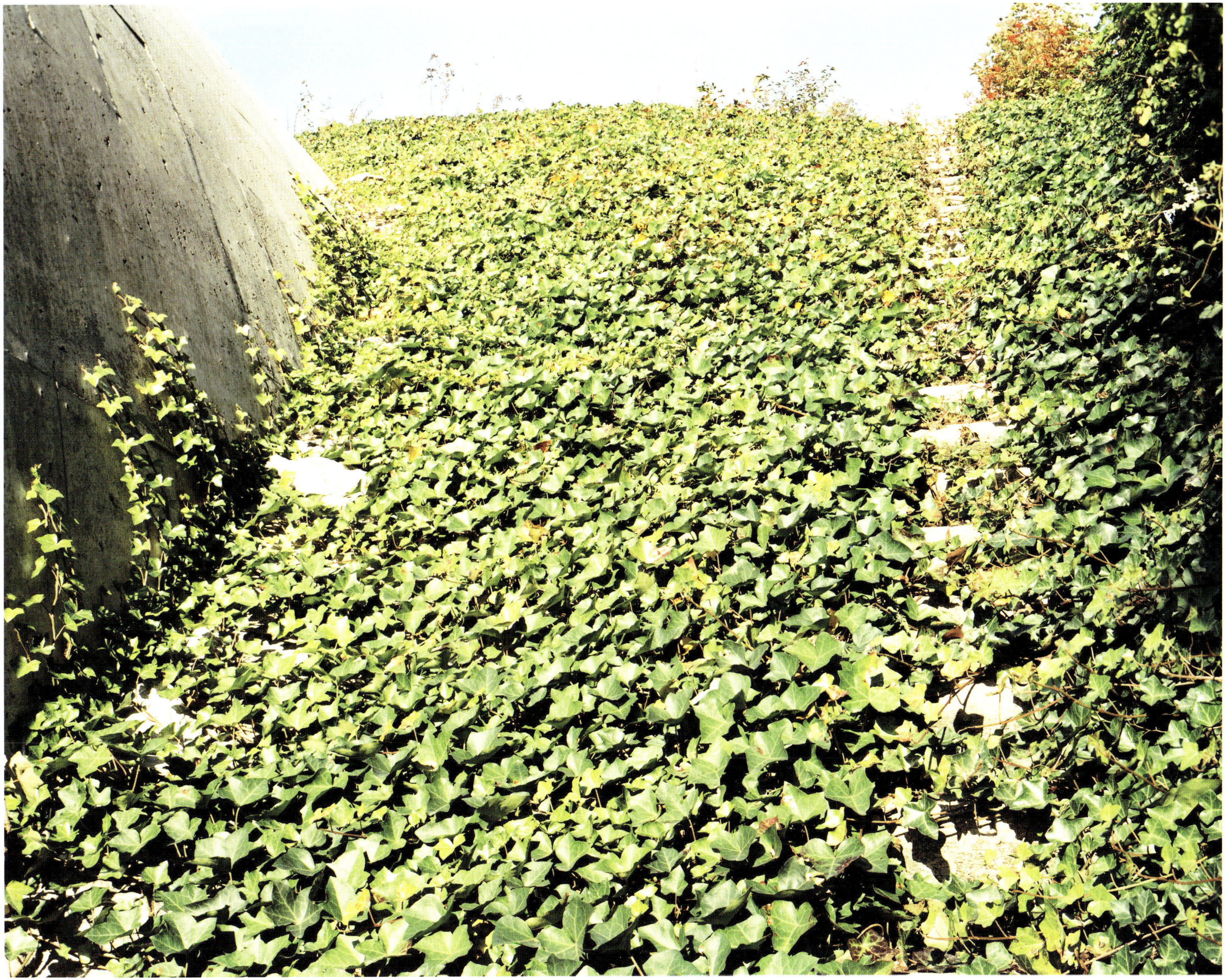

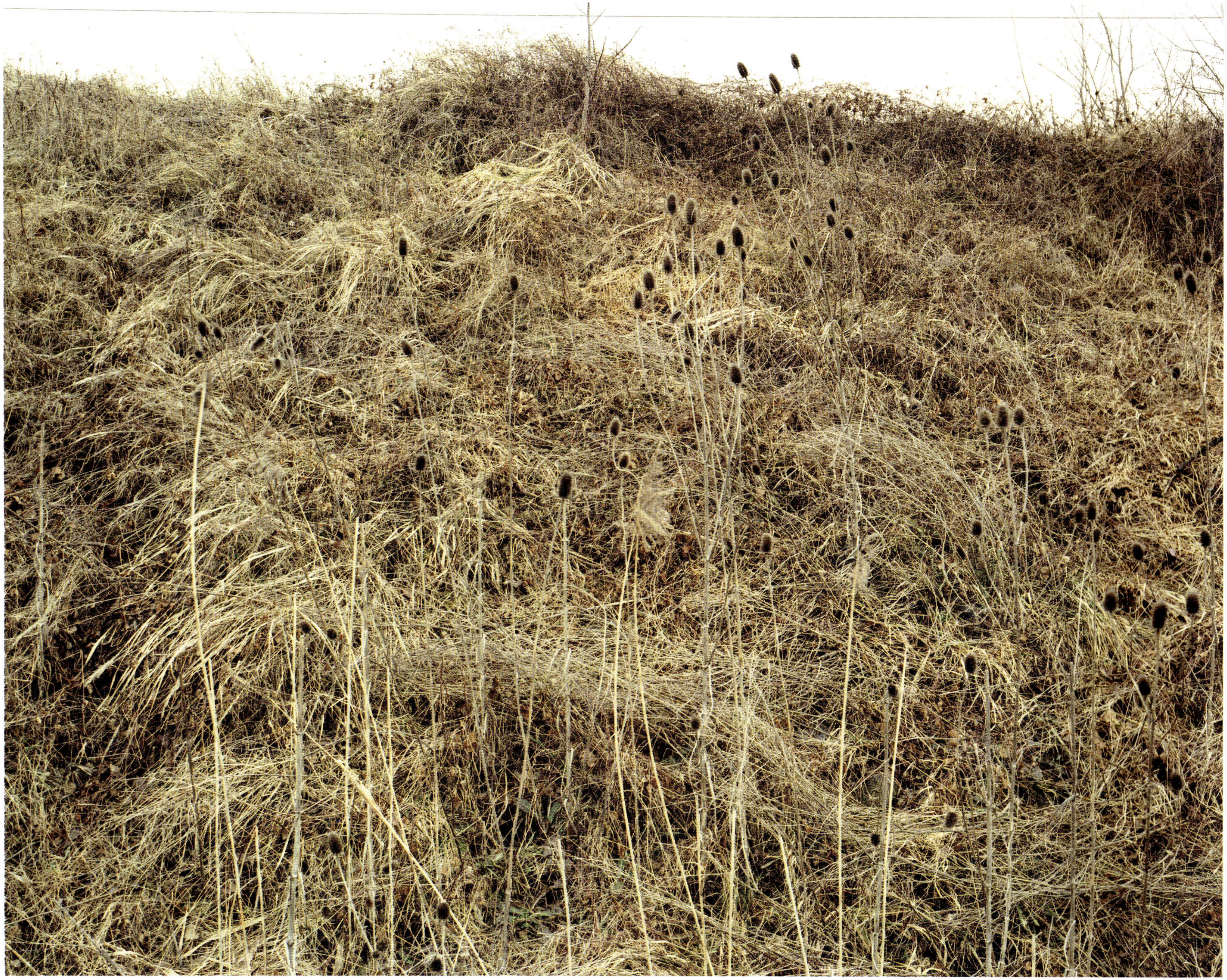

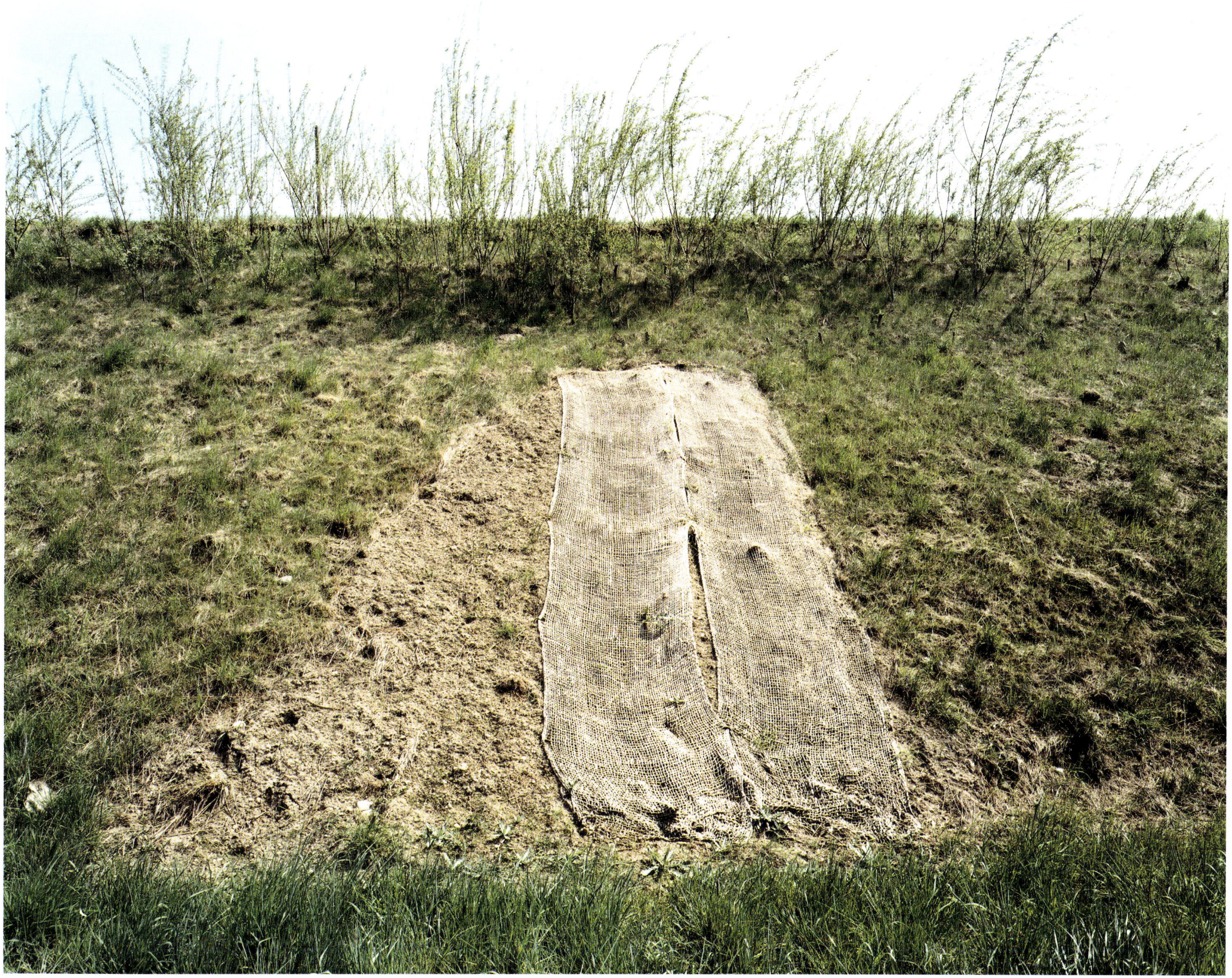

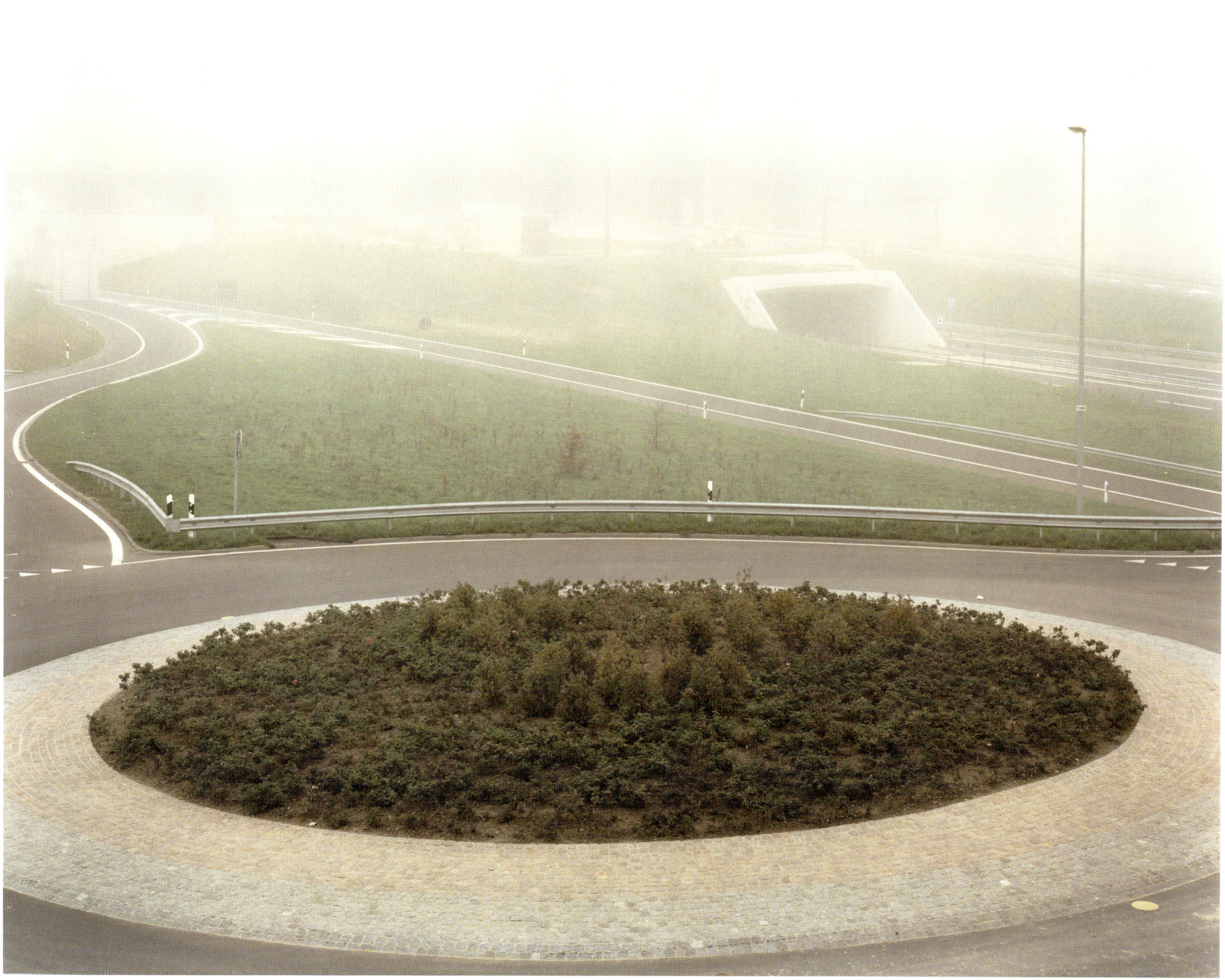

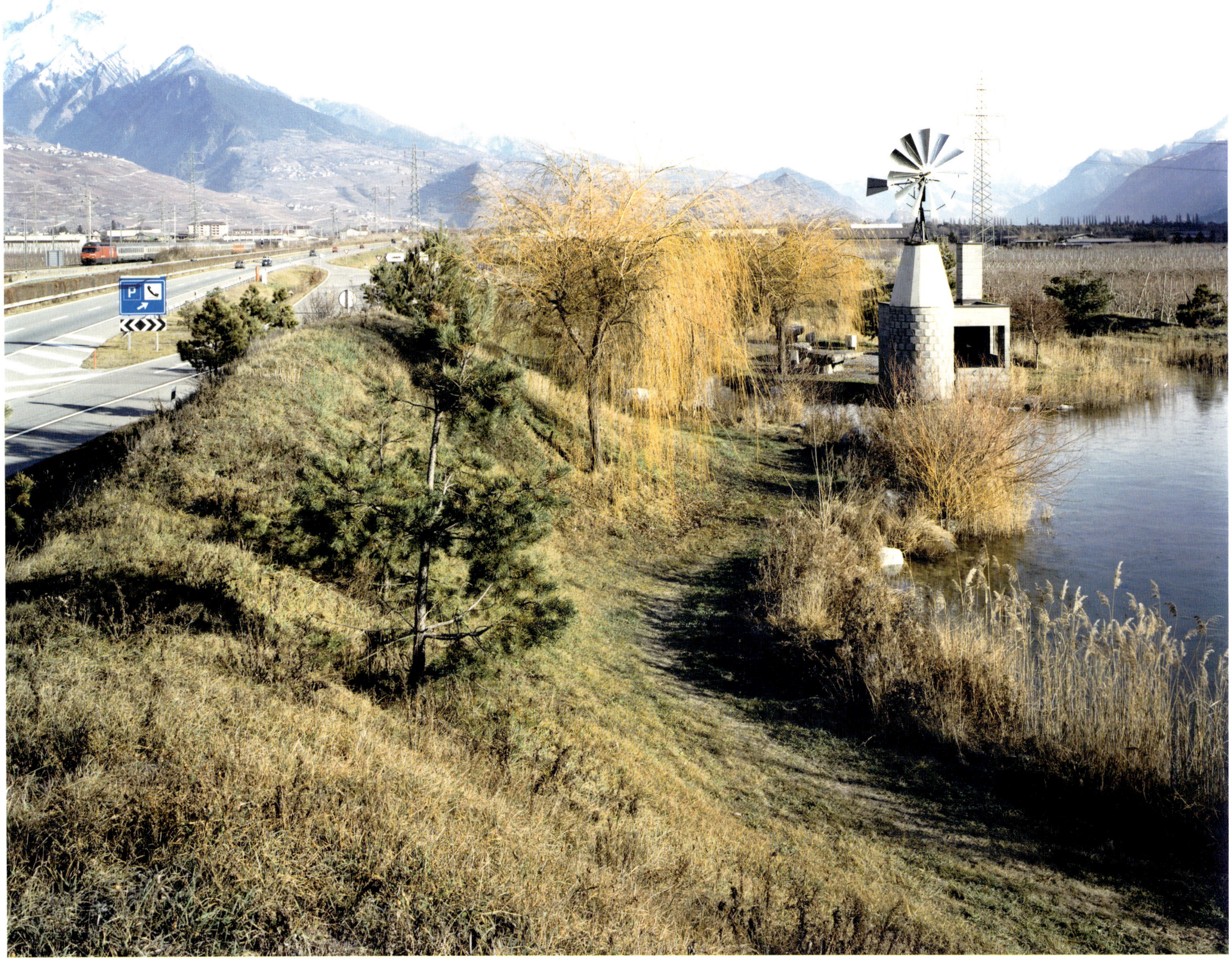
P

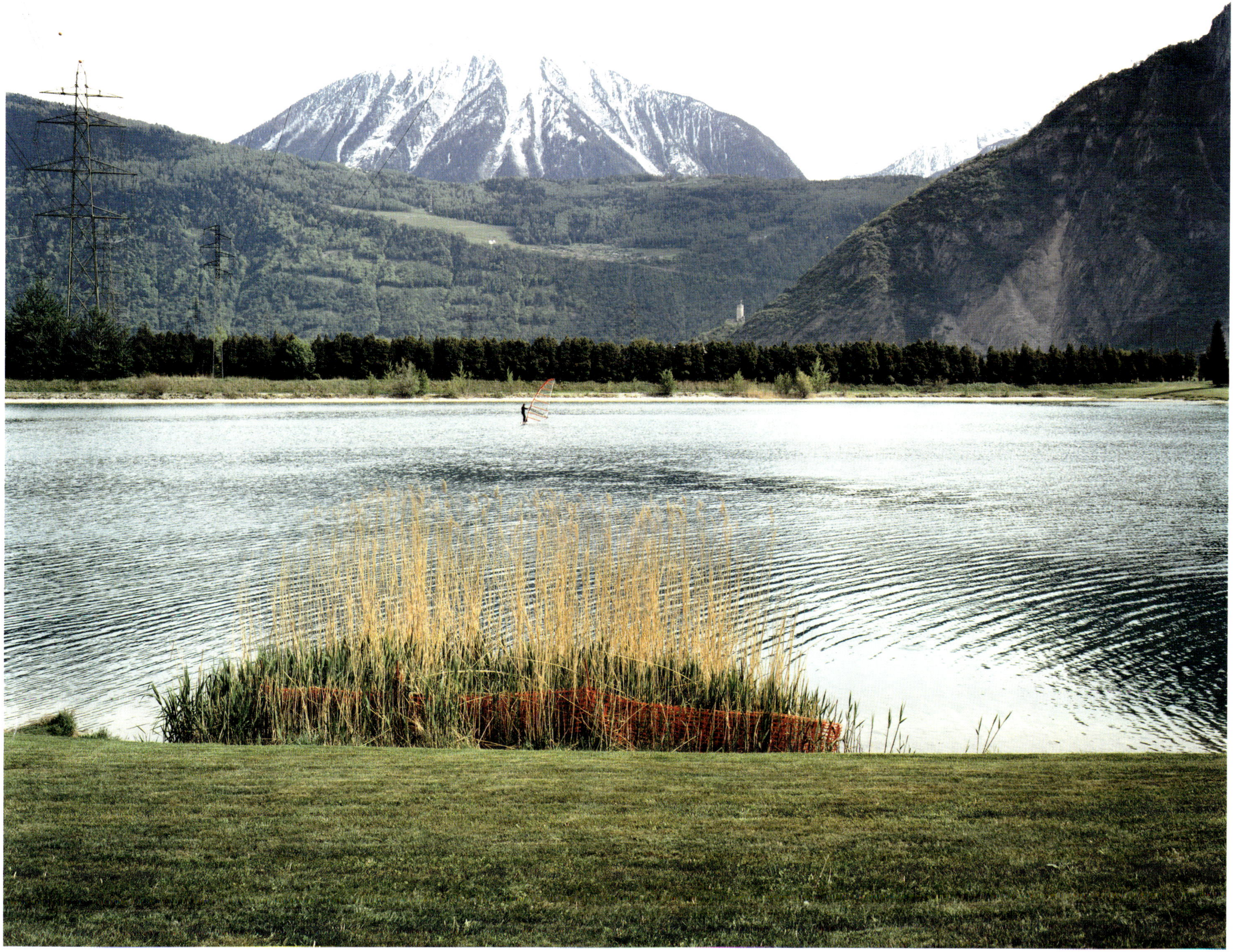

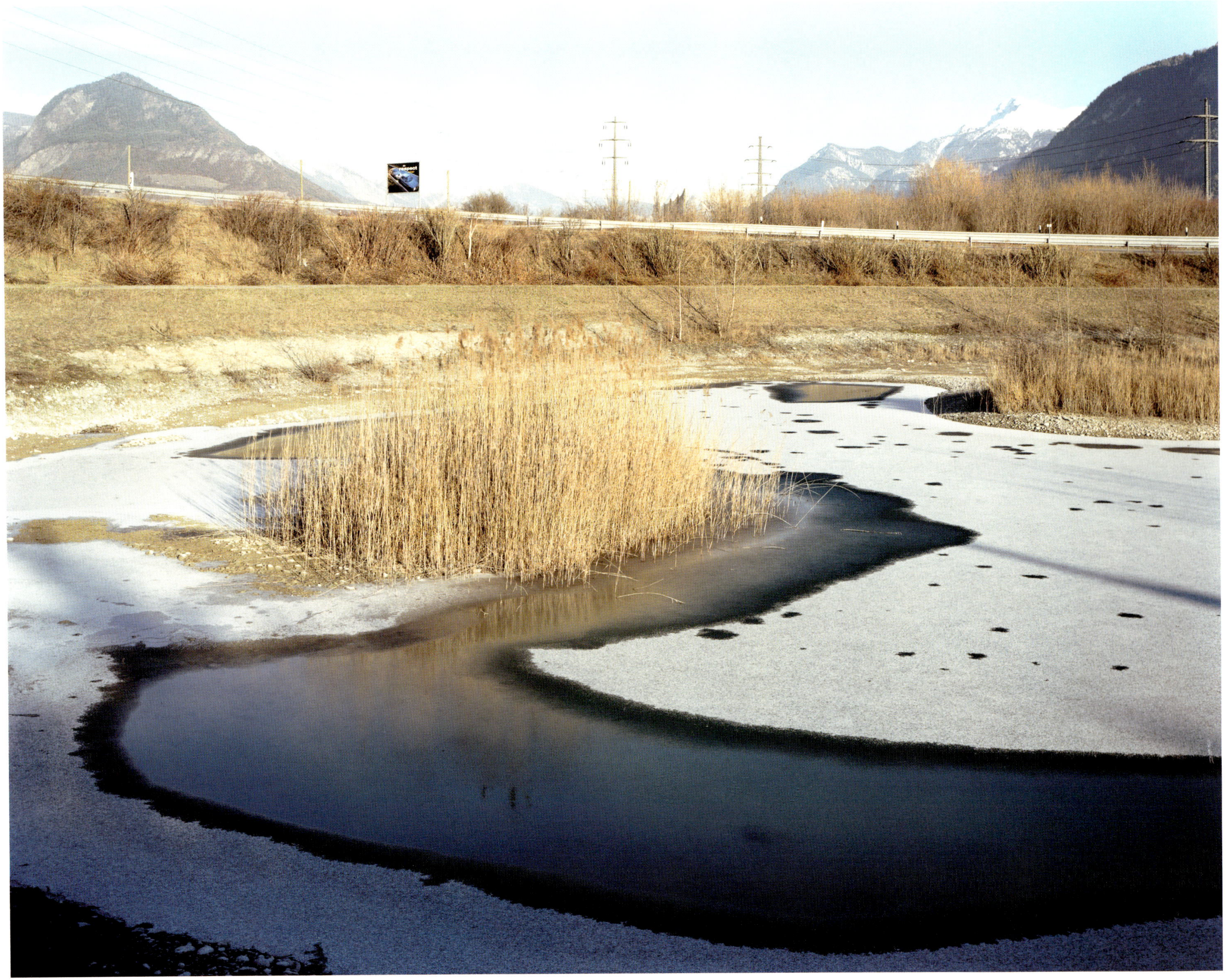

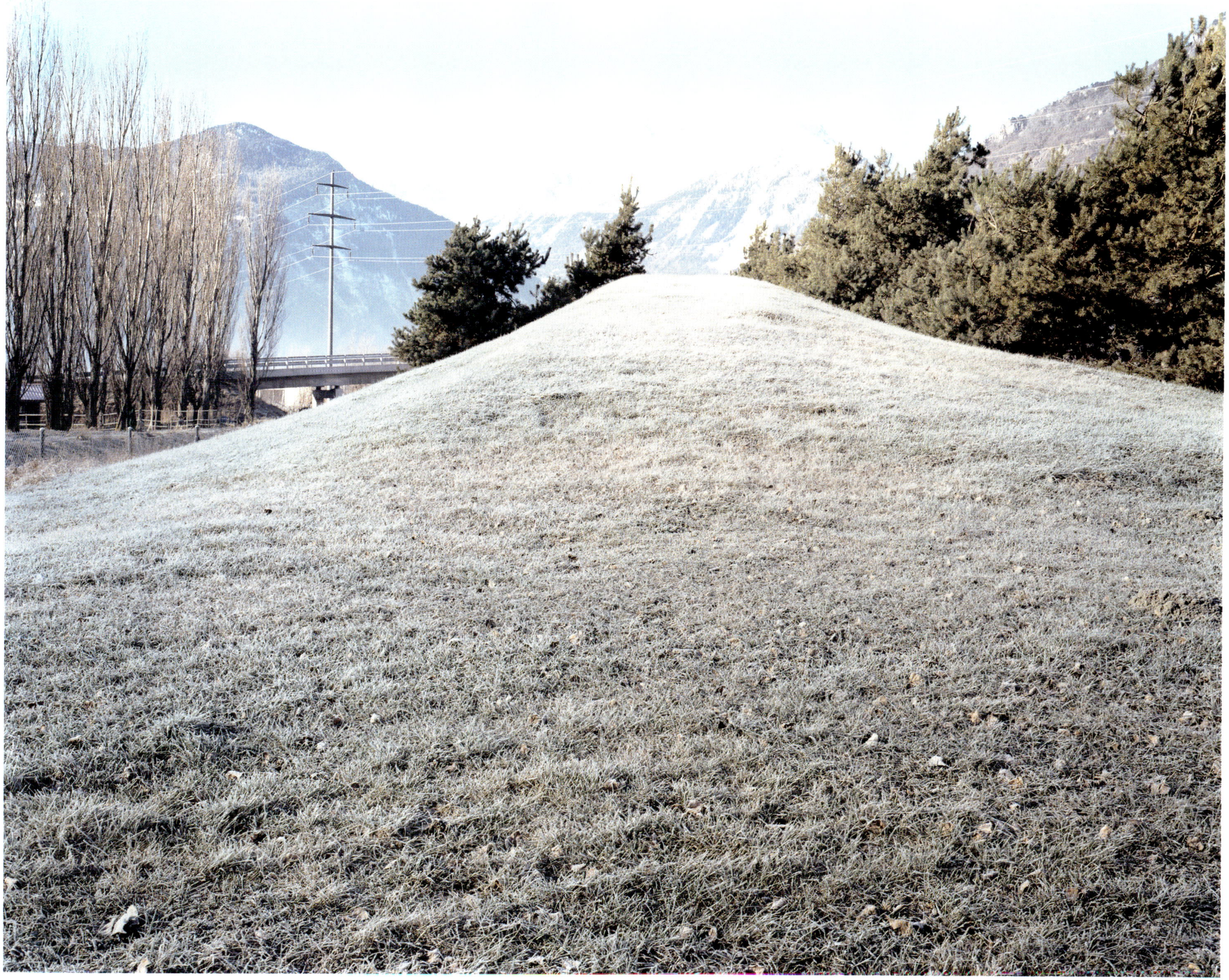

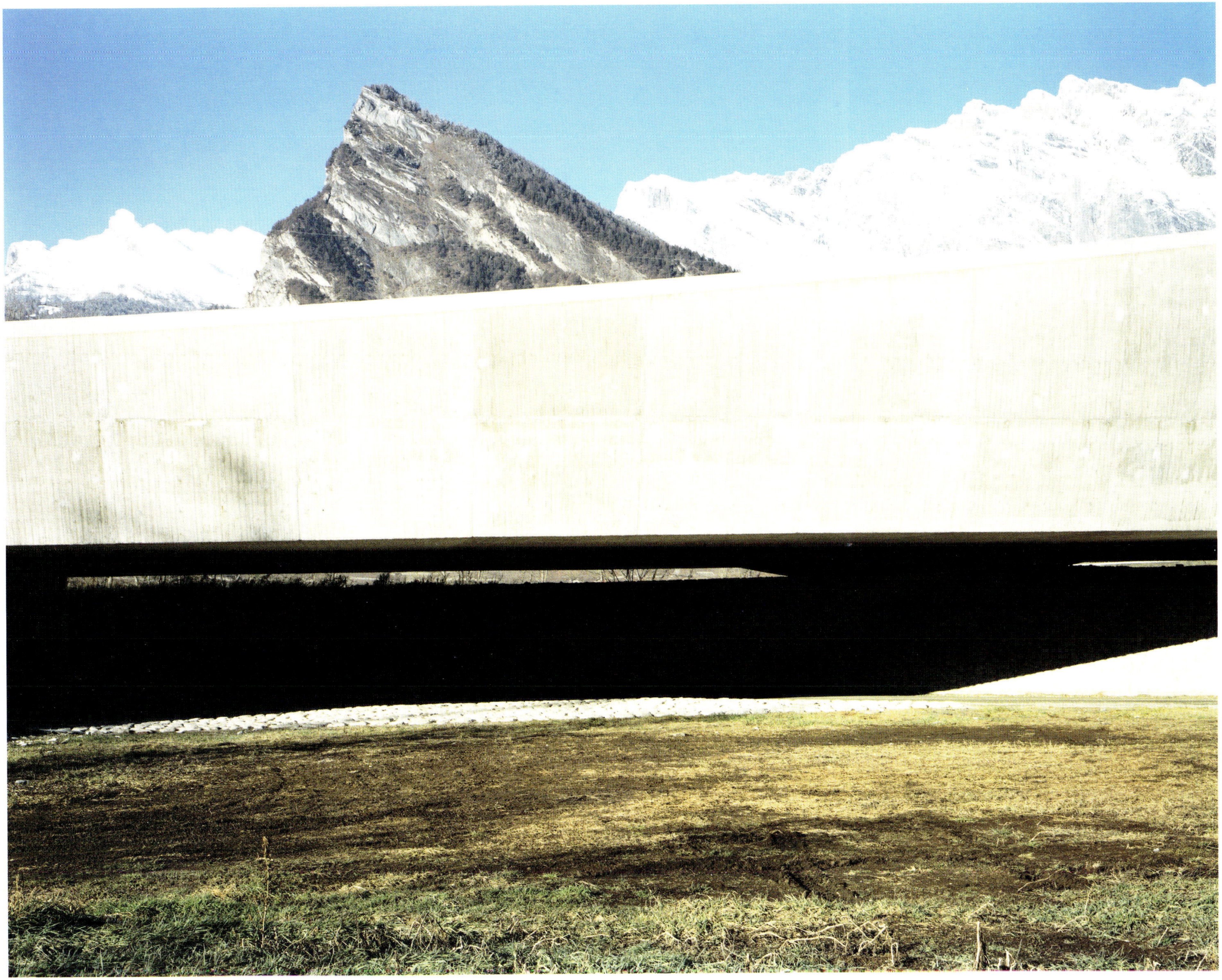

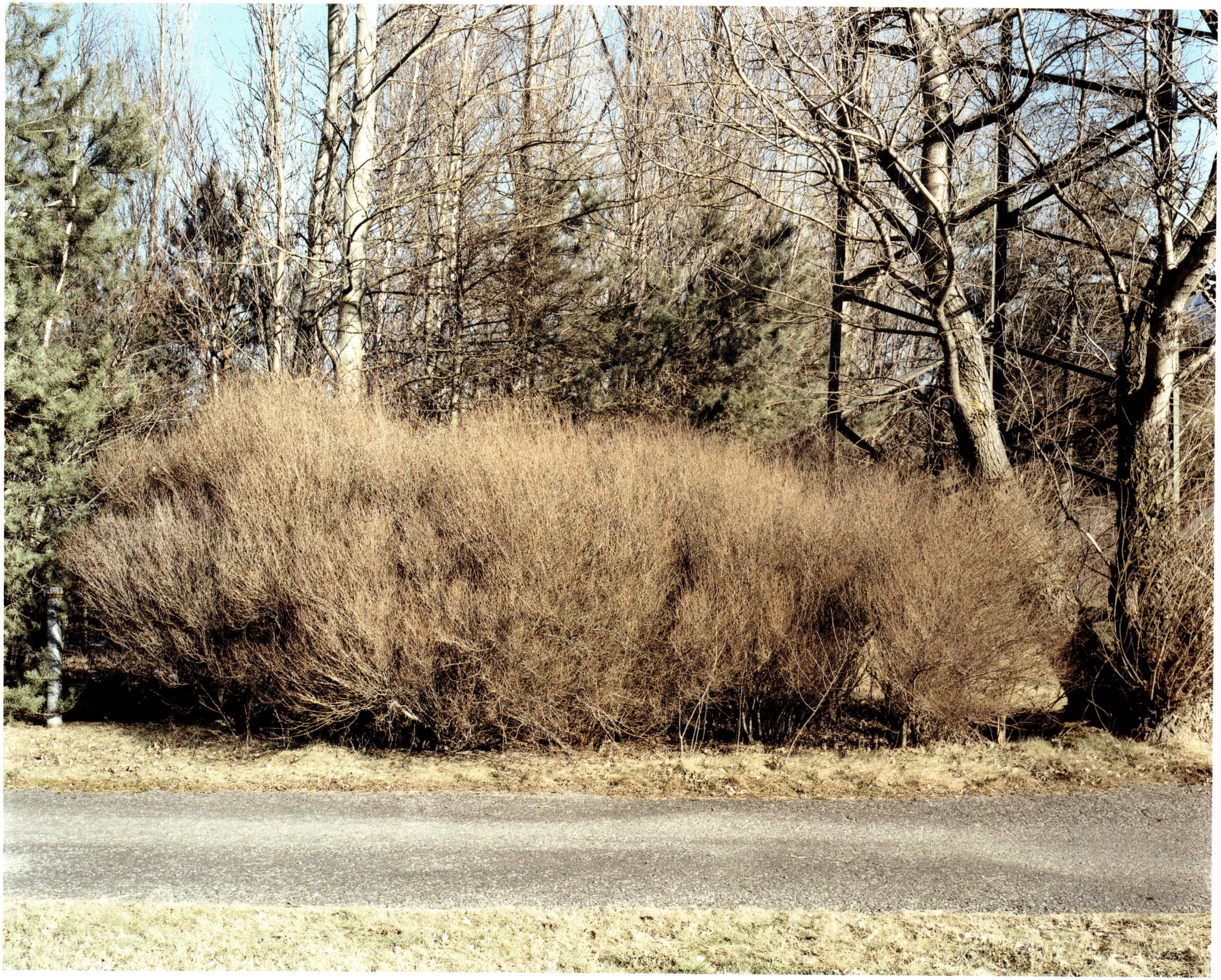

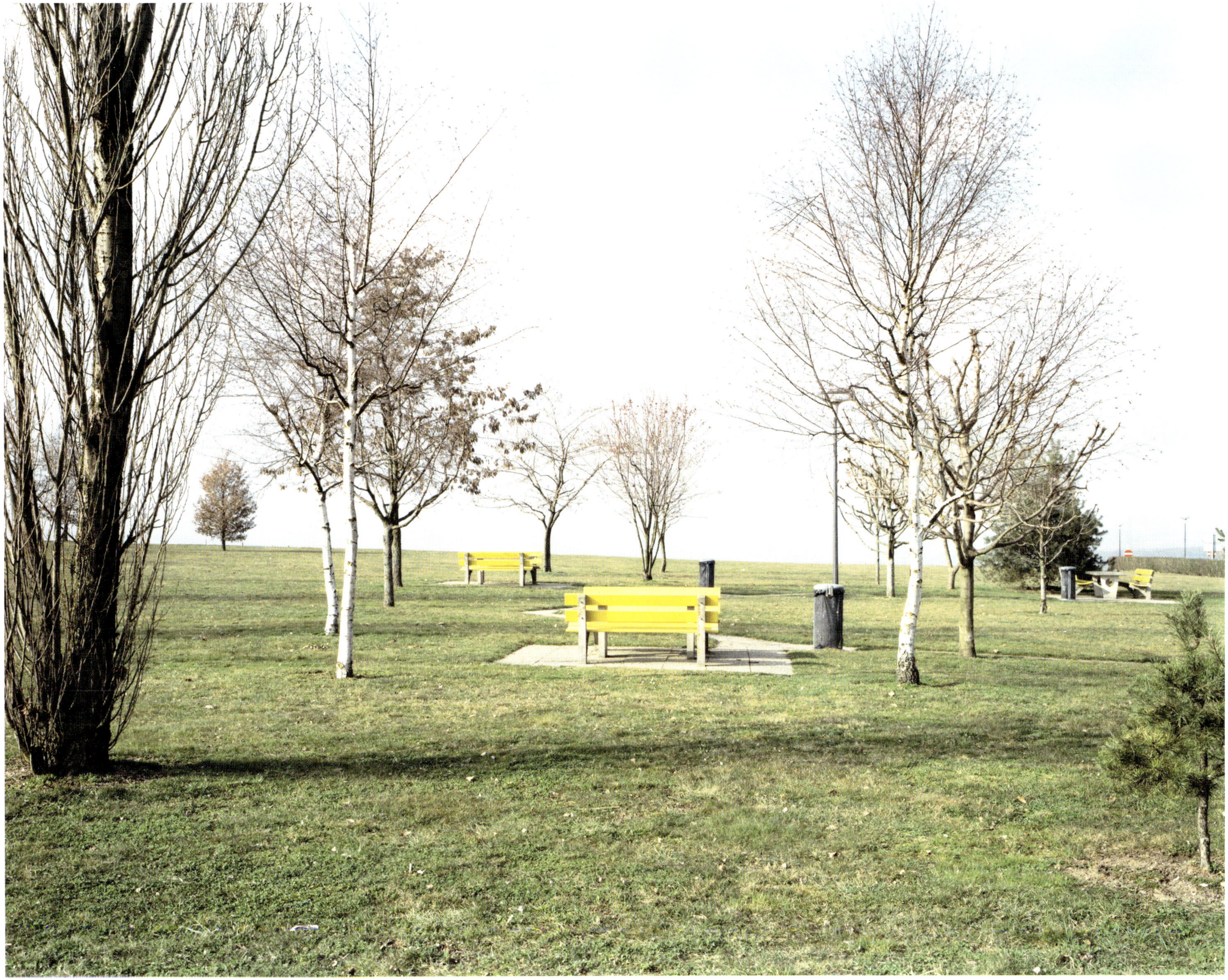

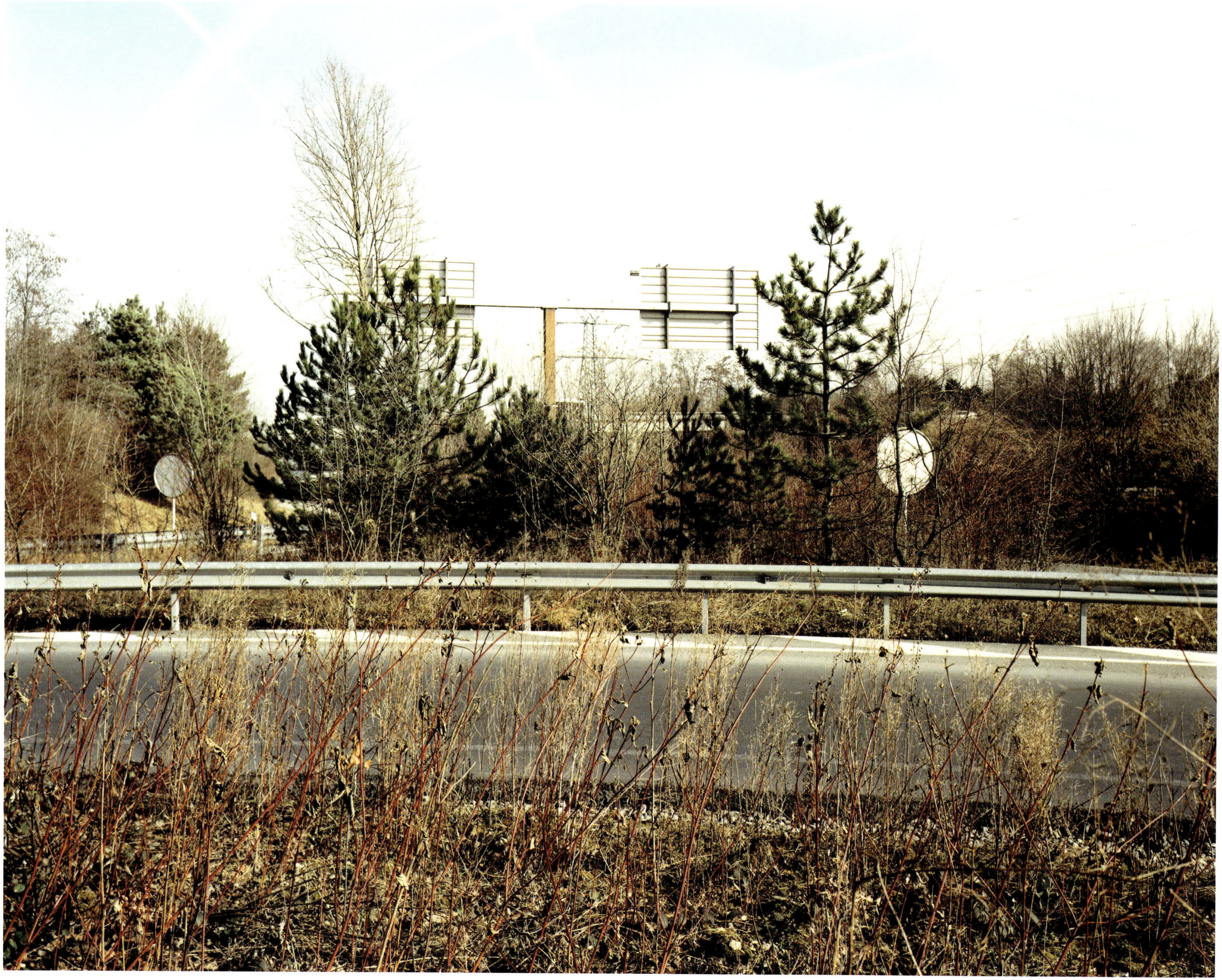

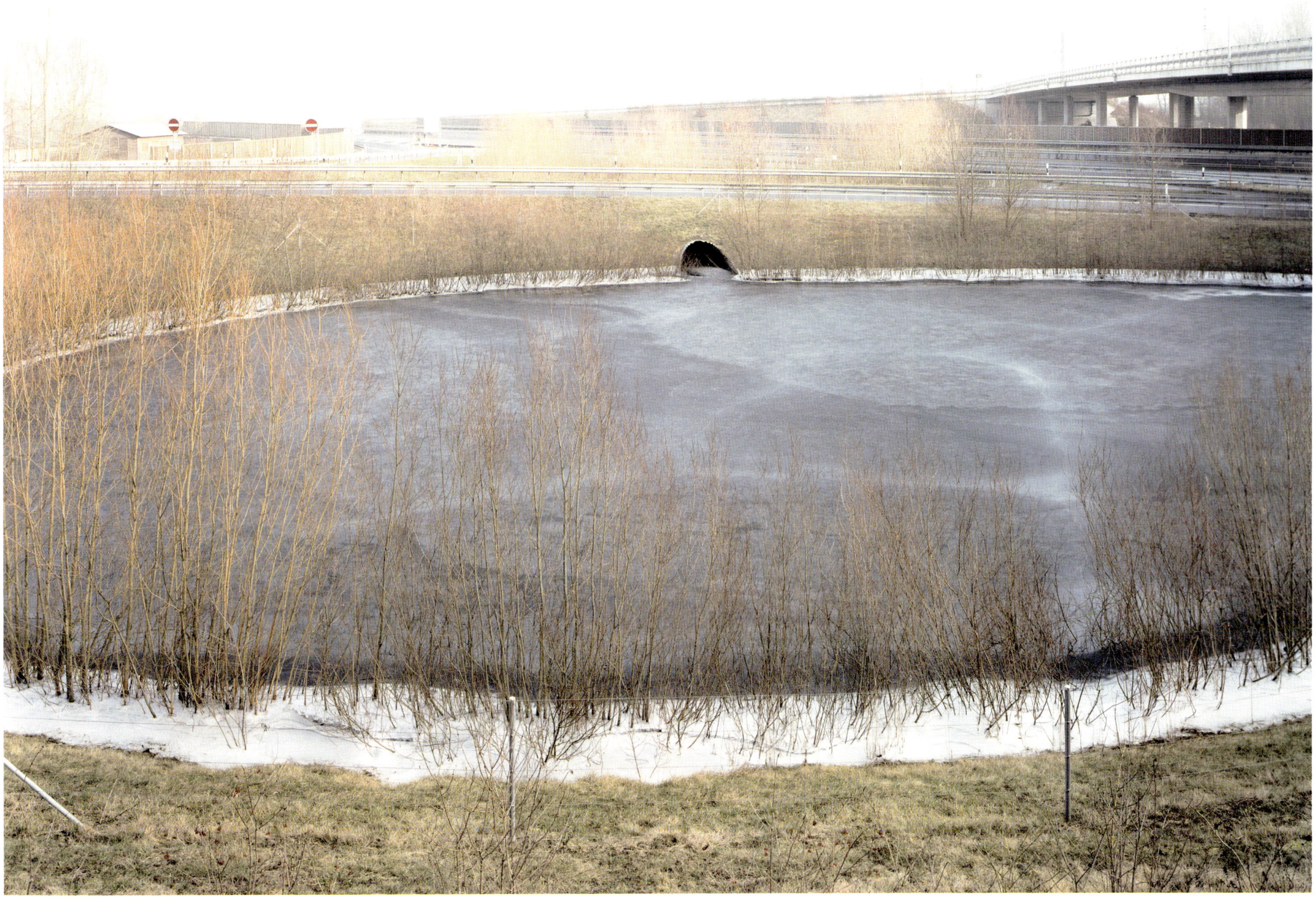

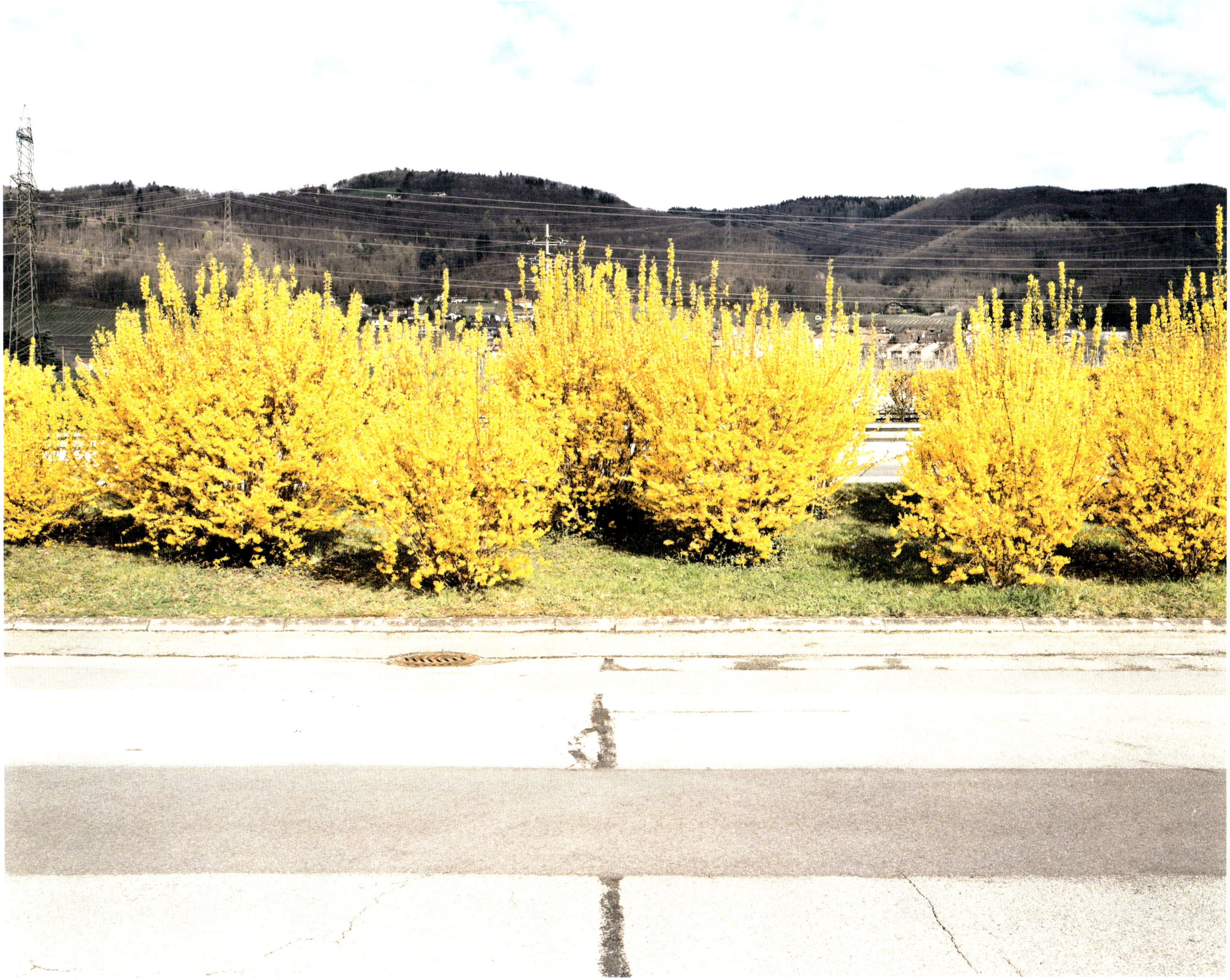

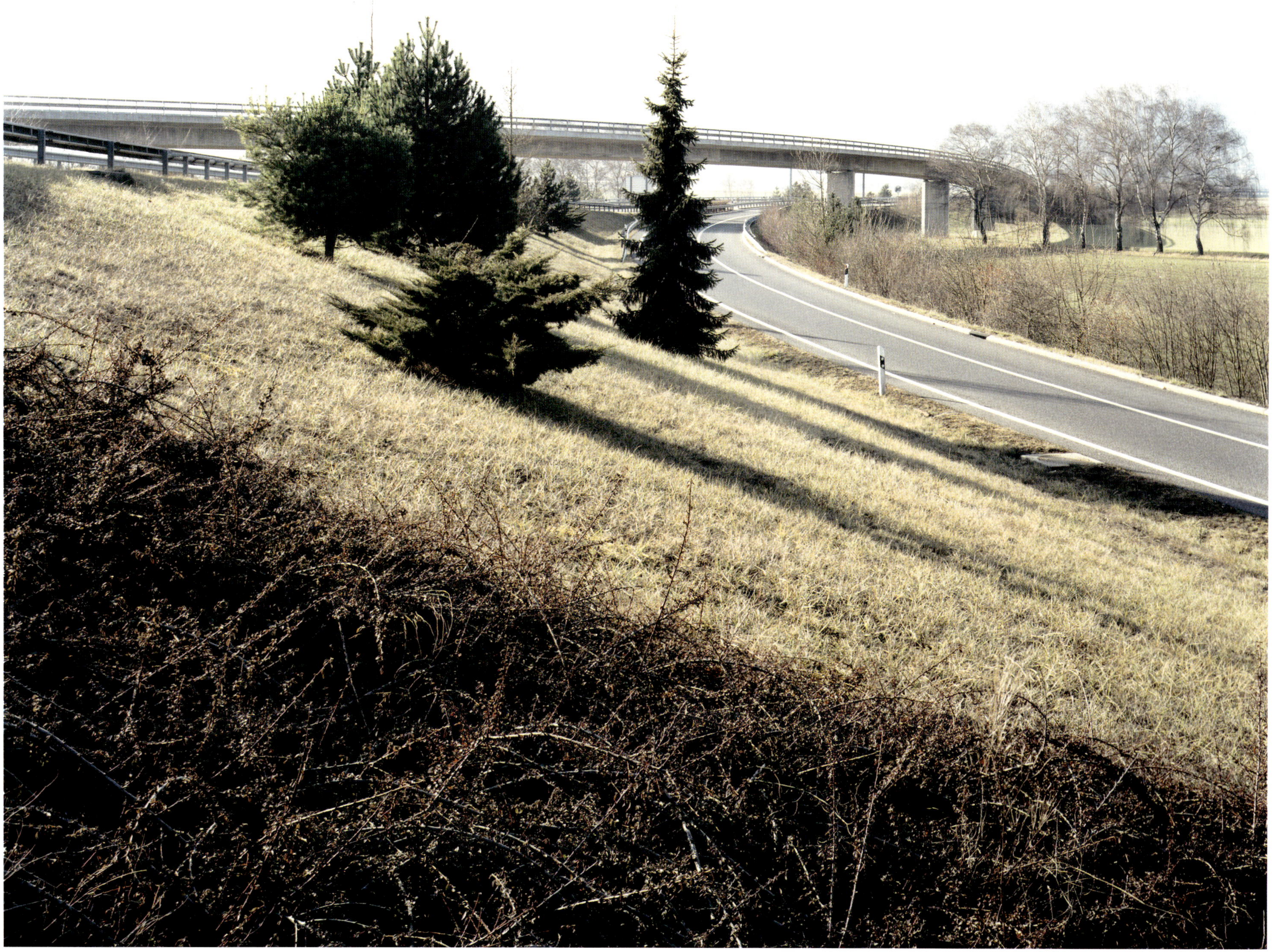

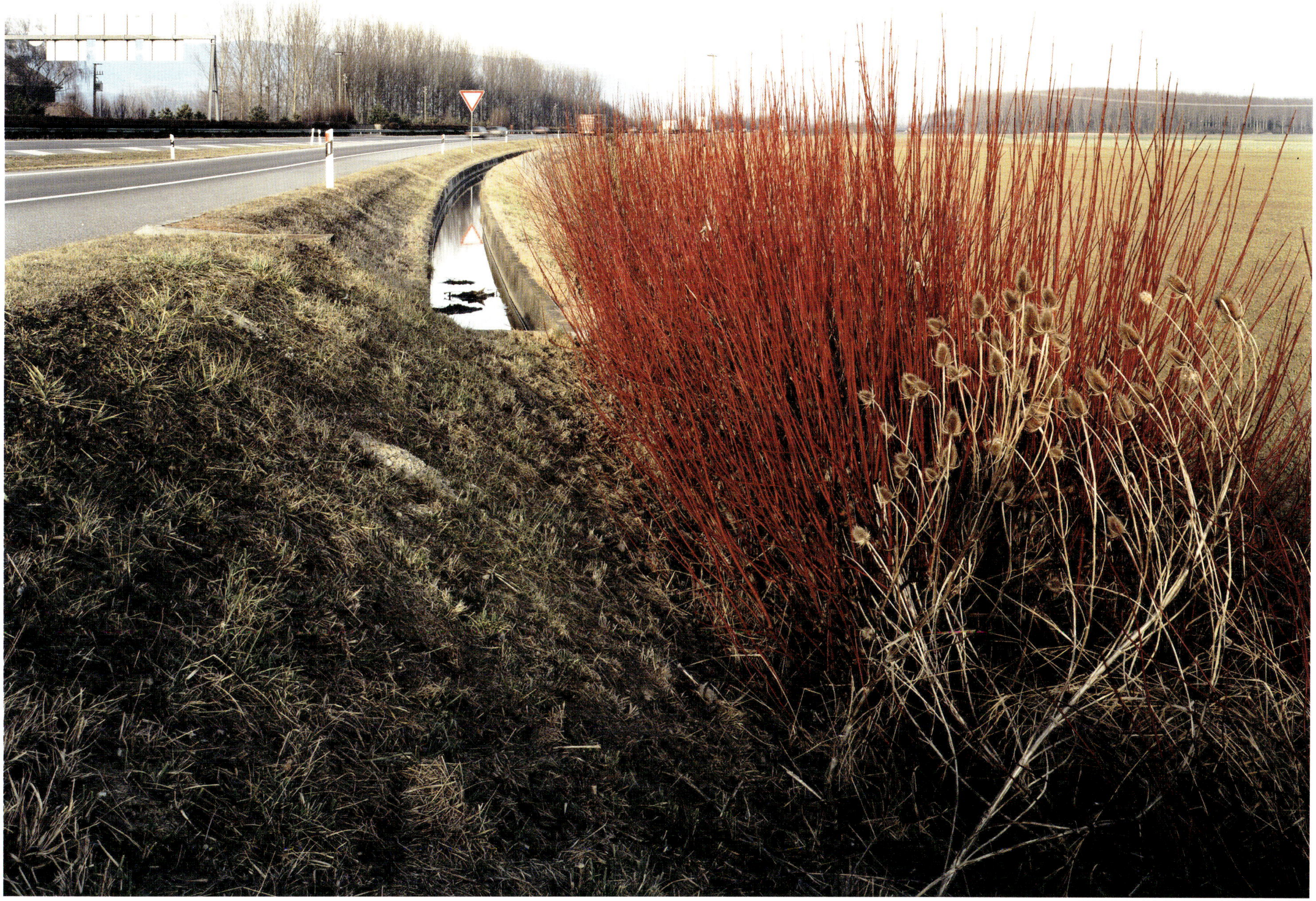

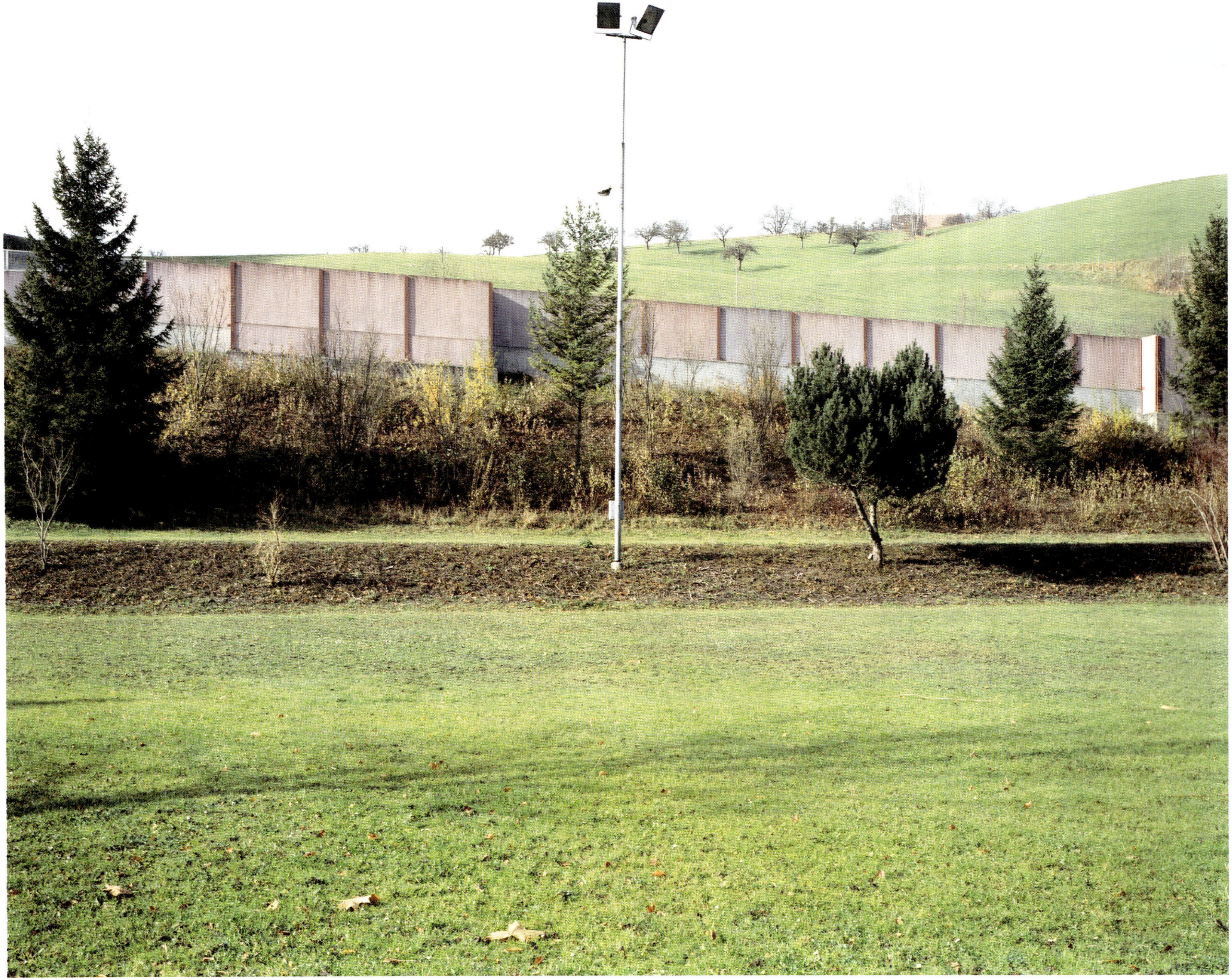

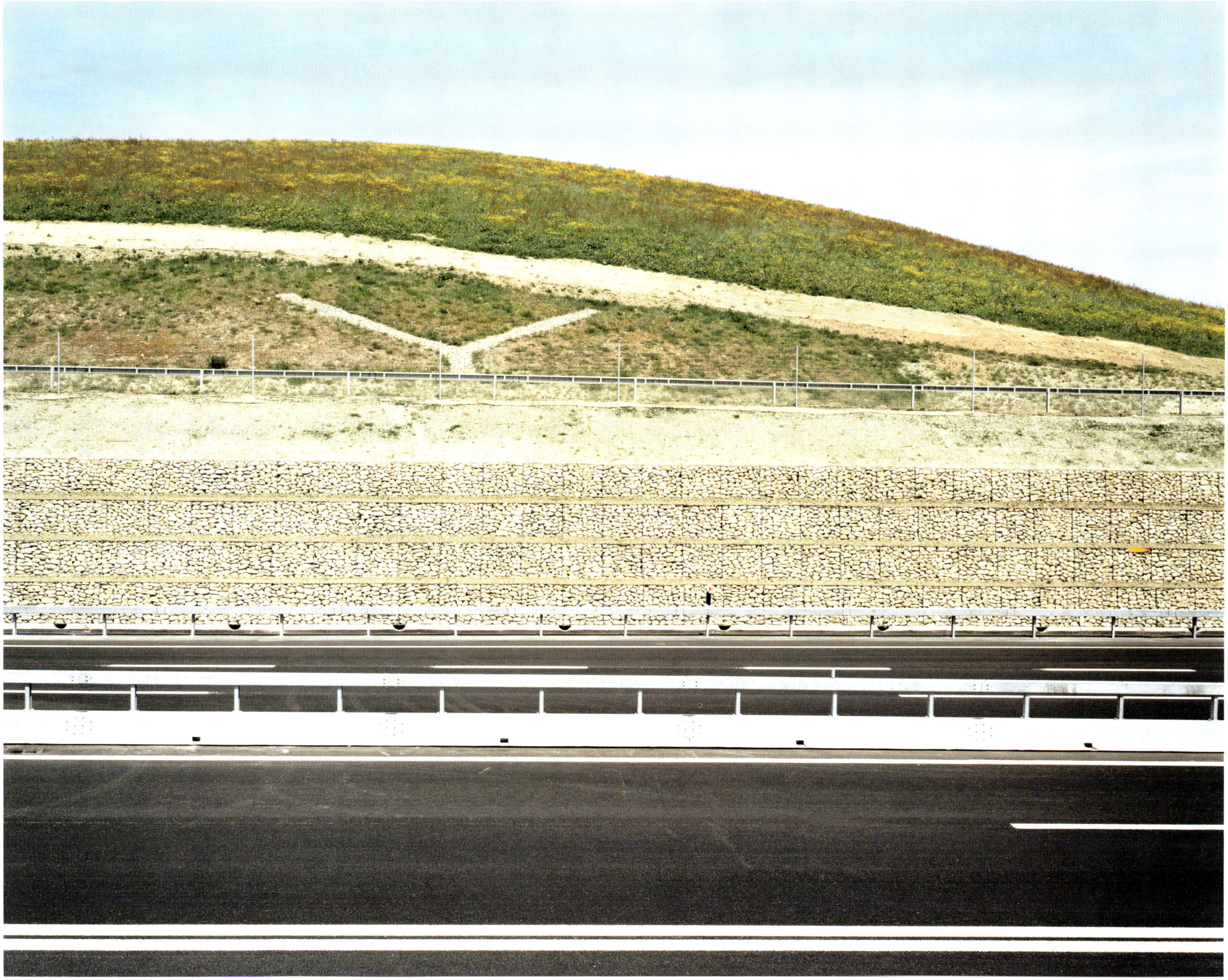

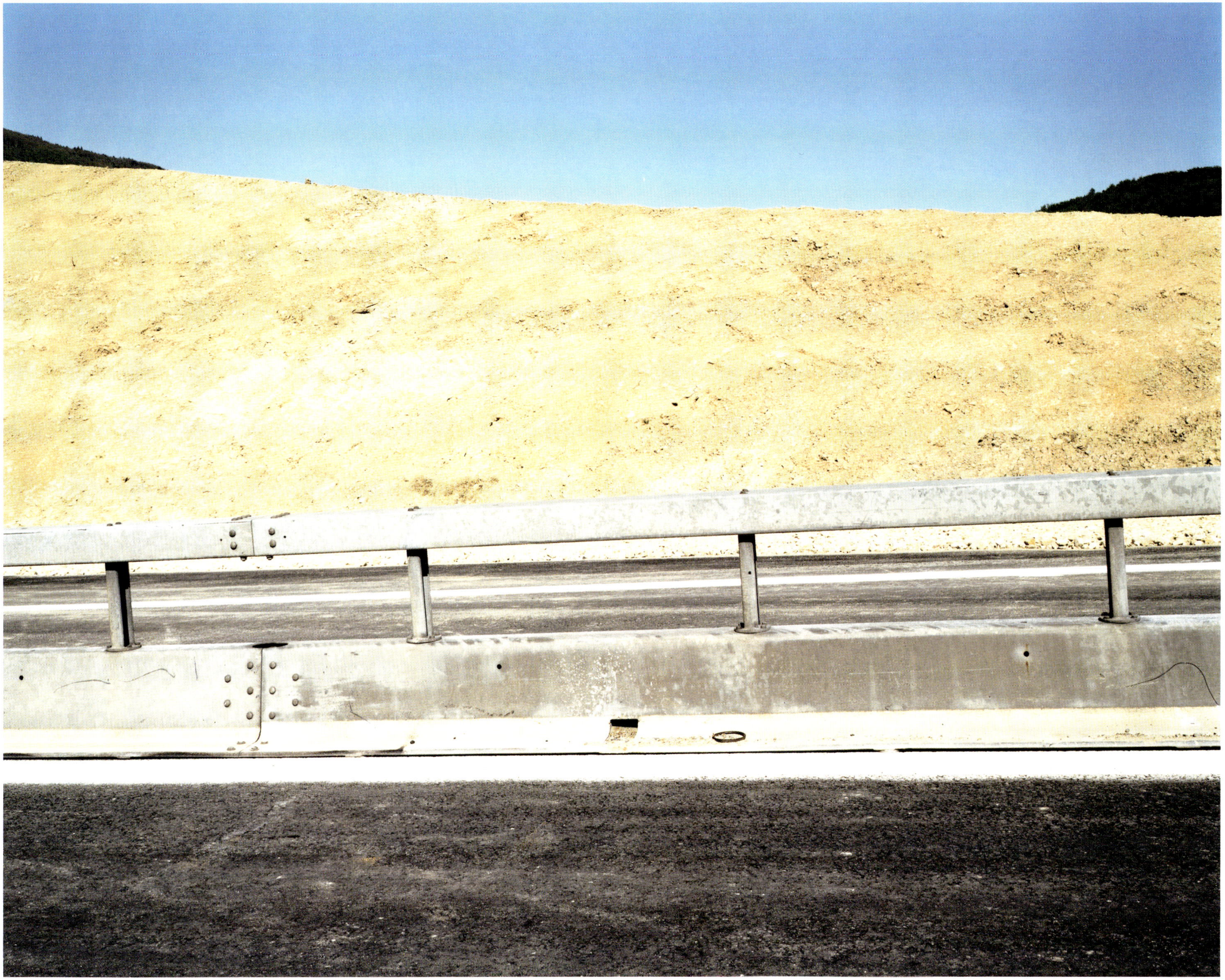

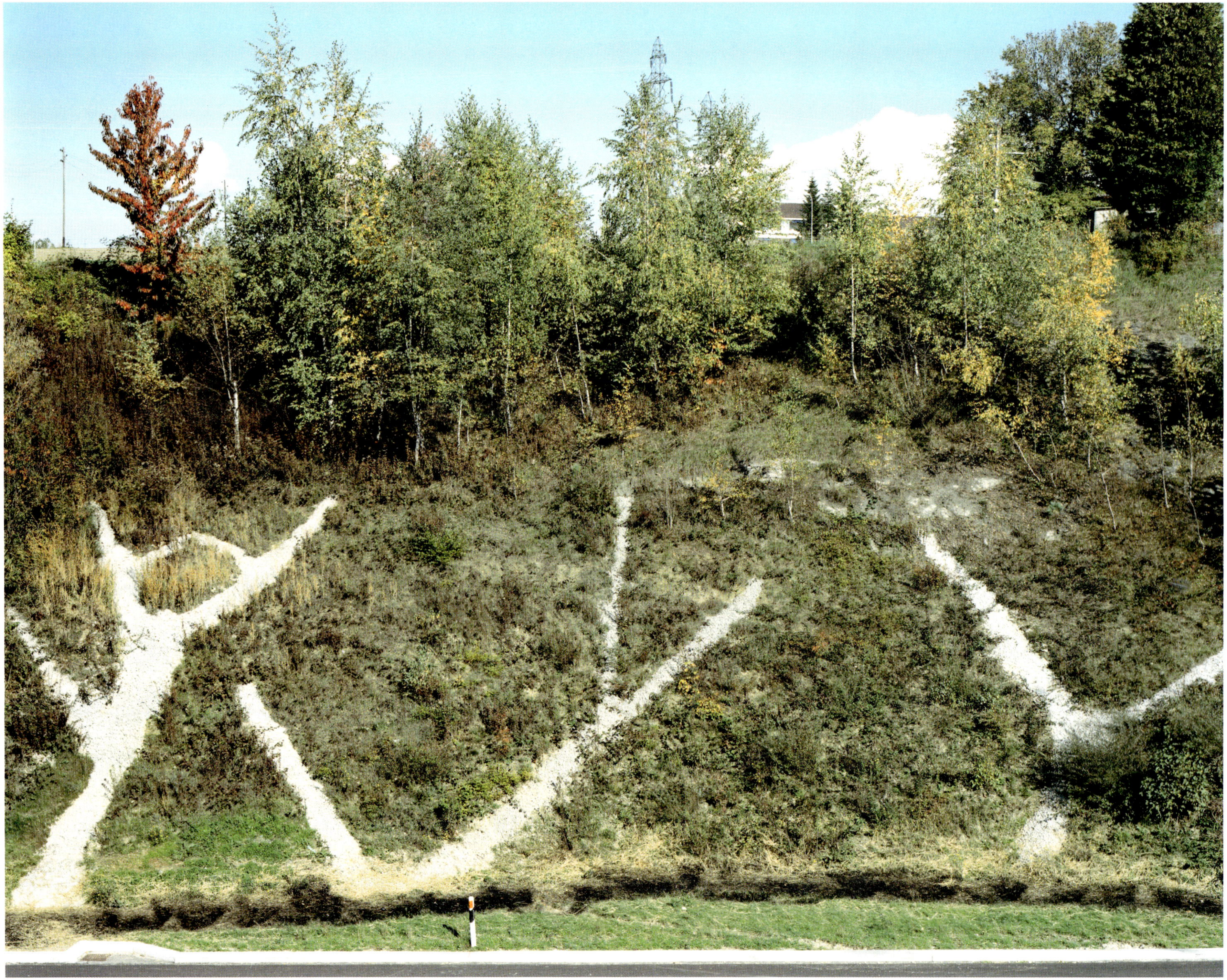

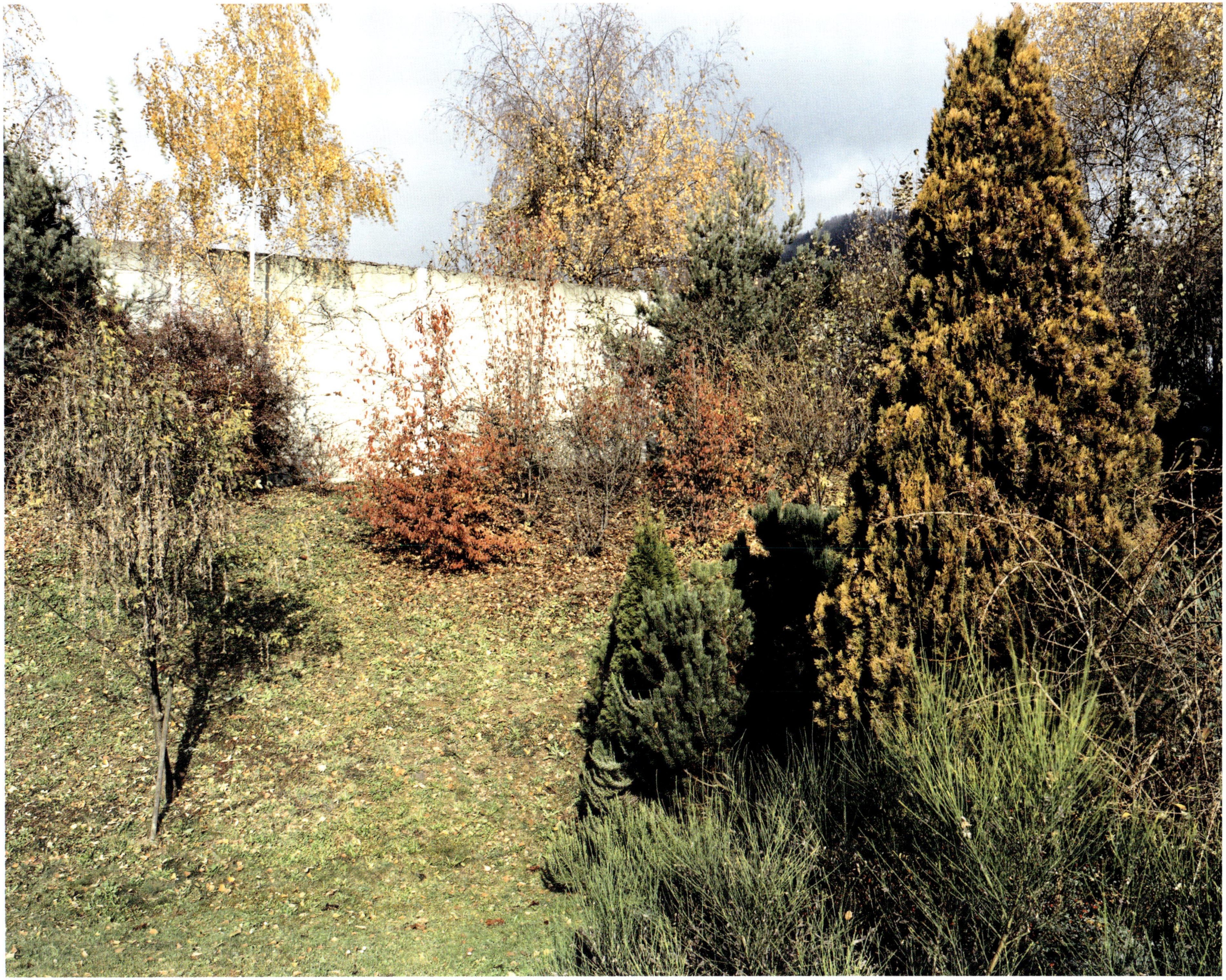

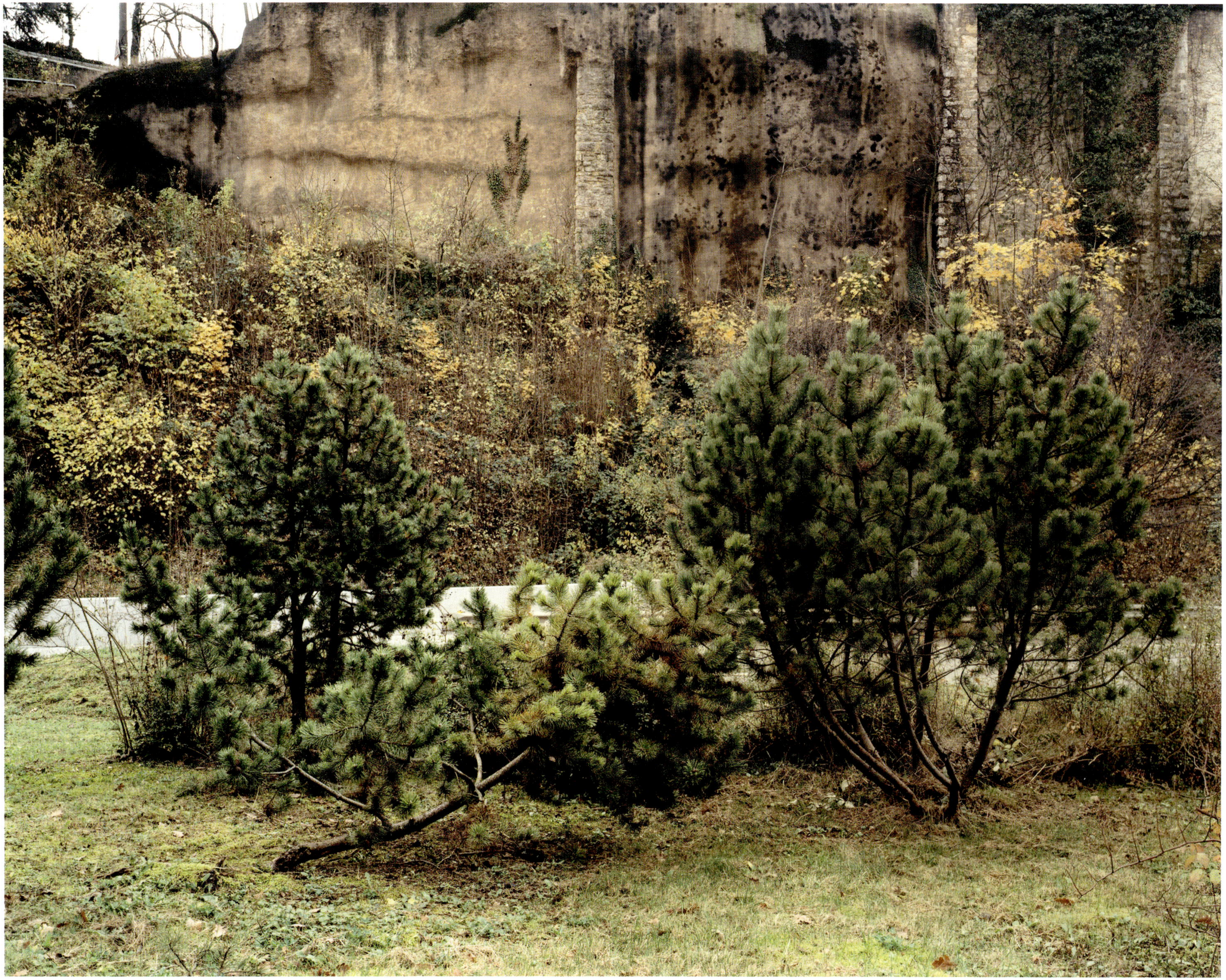

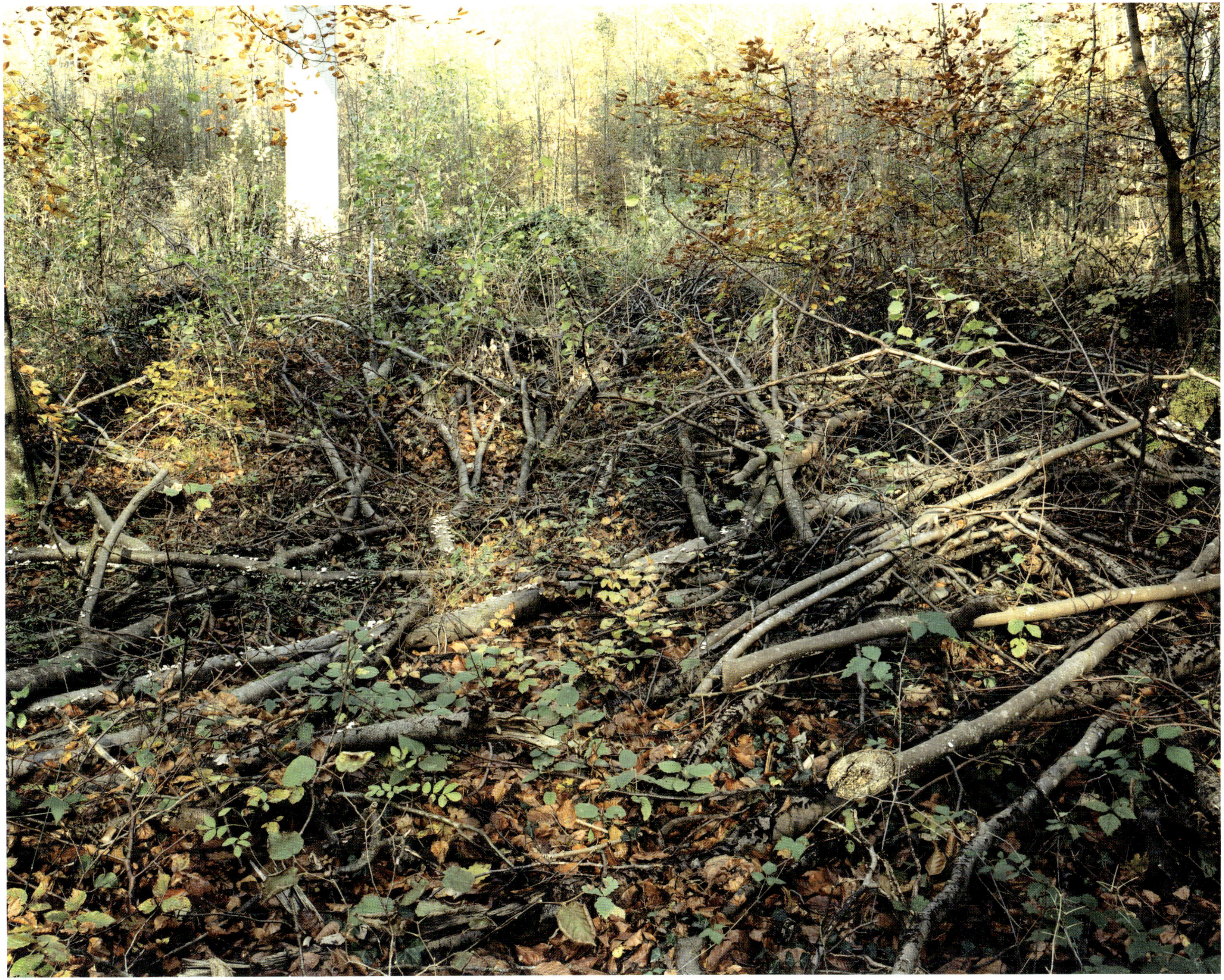

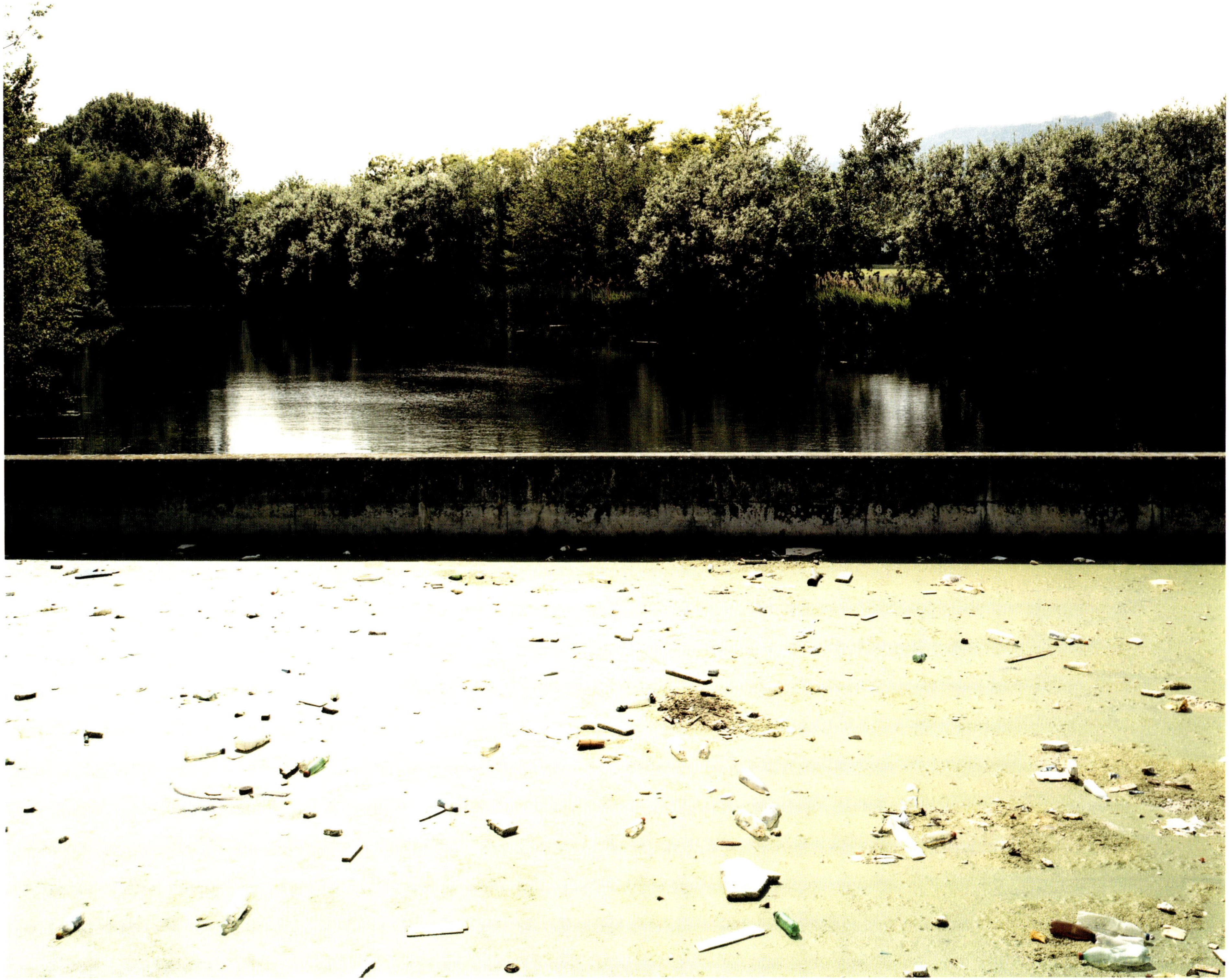

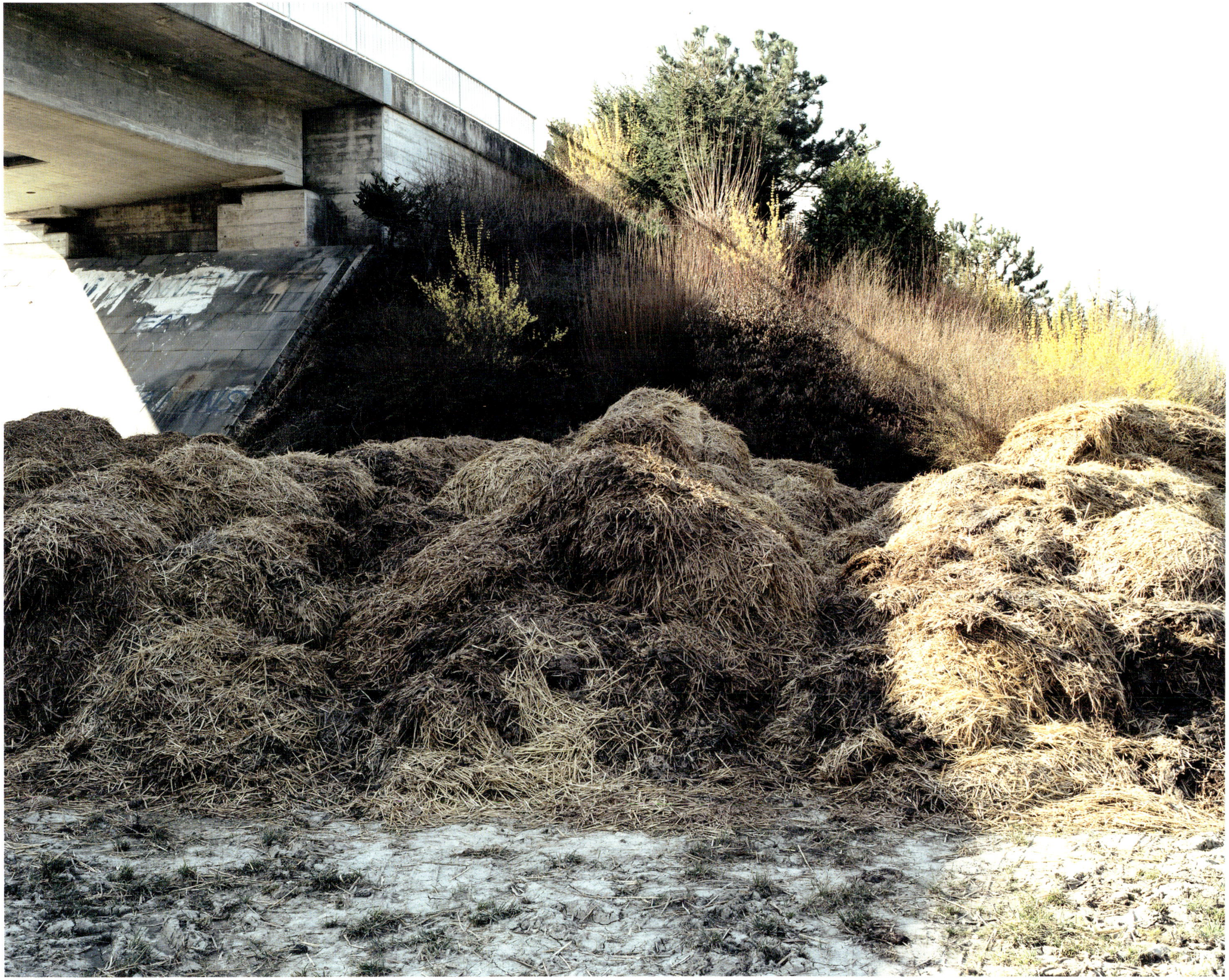

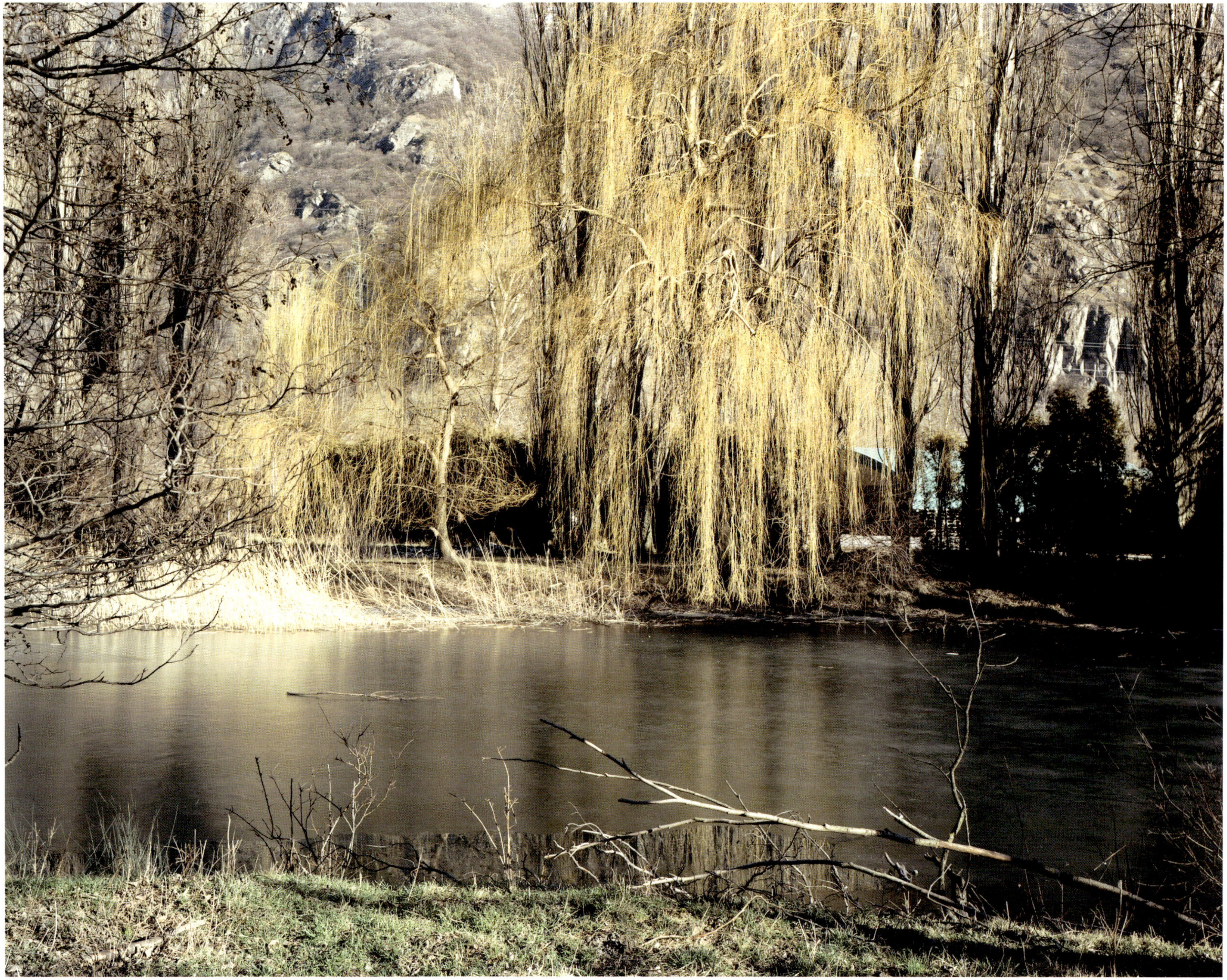

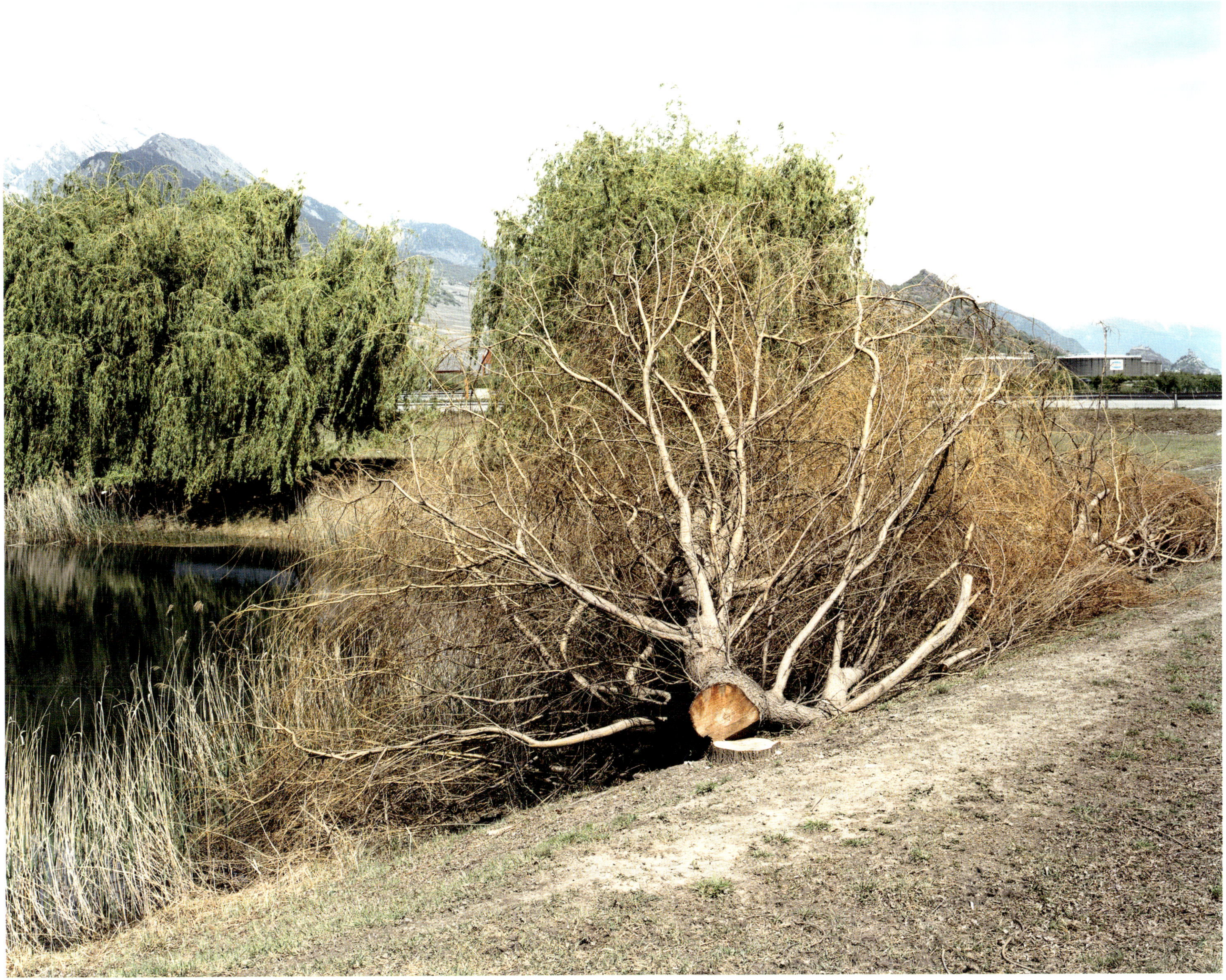

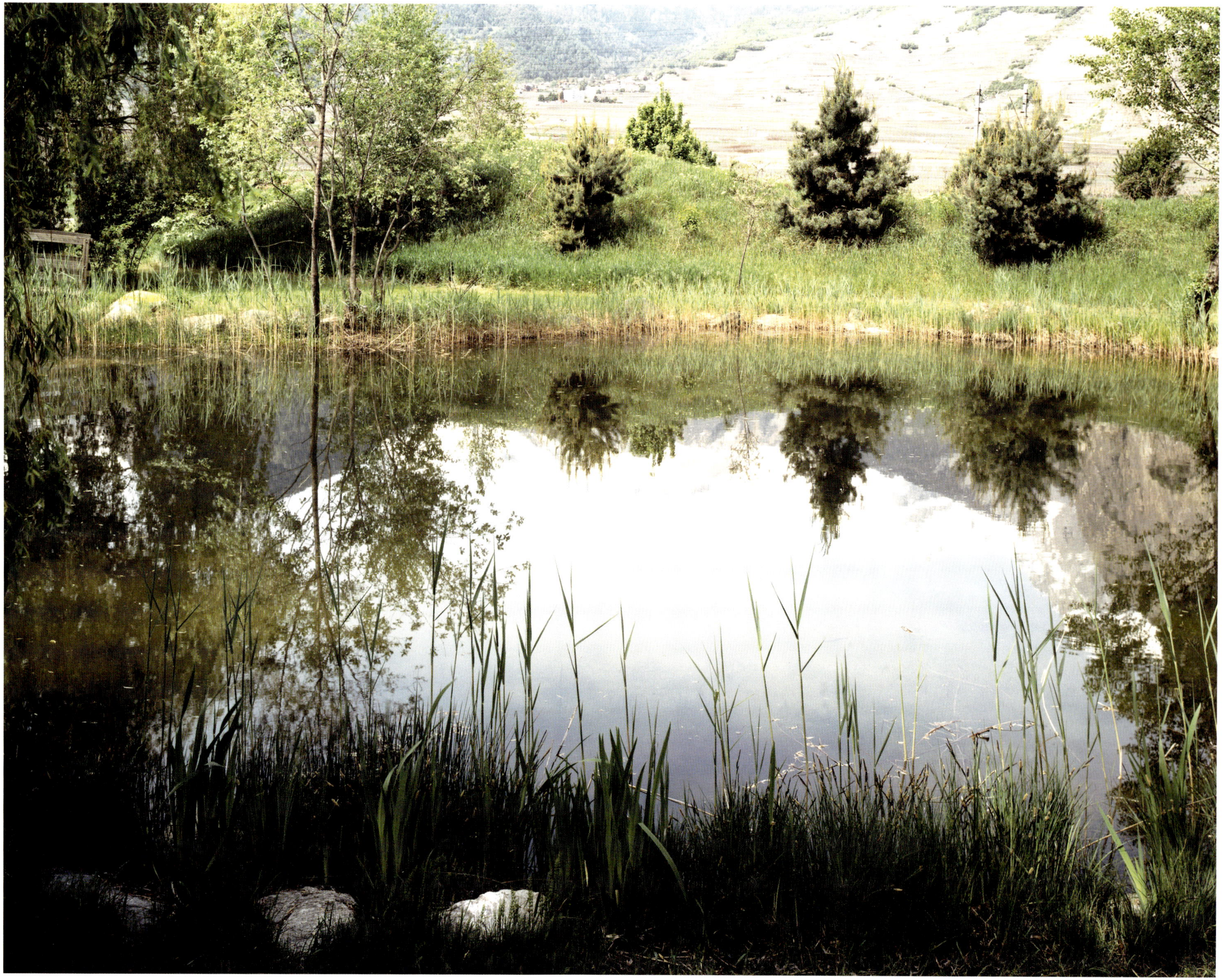

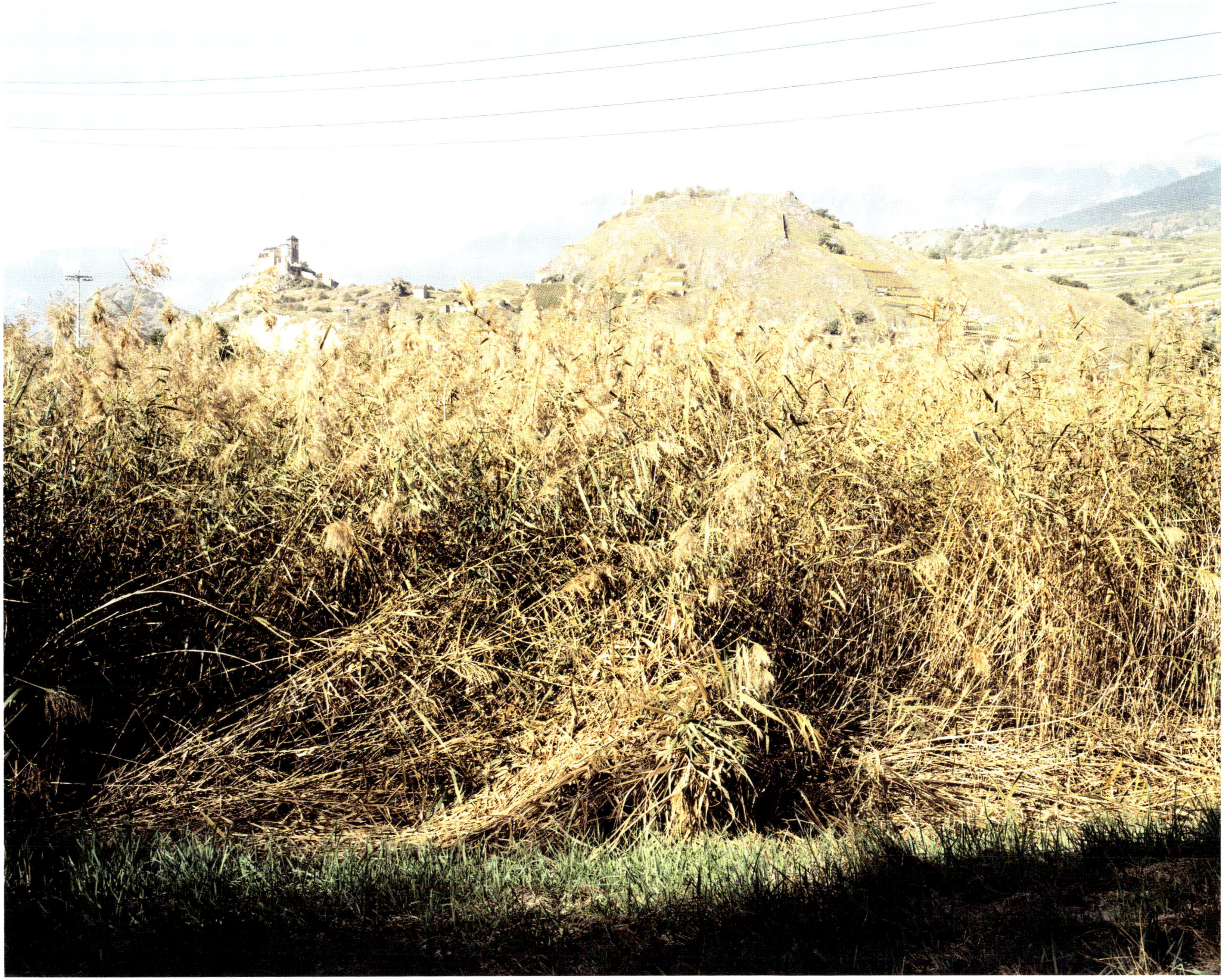

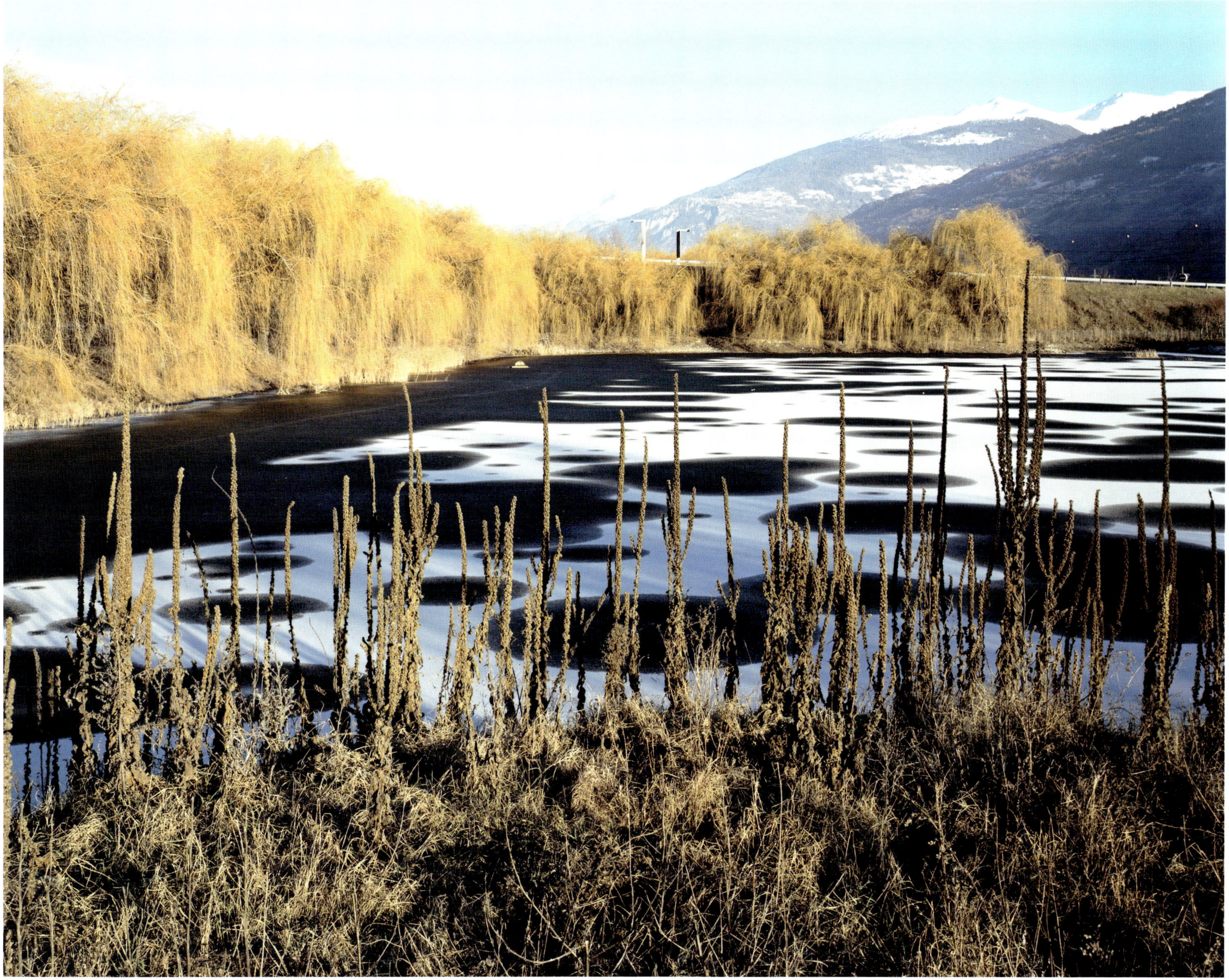

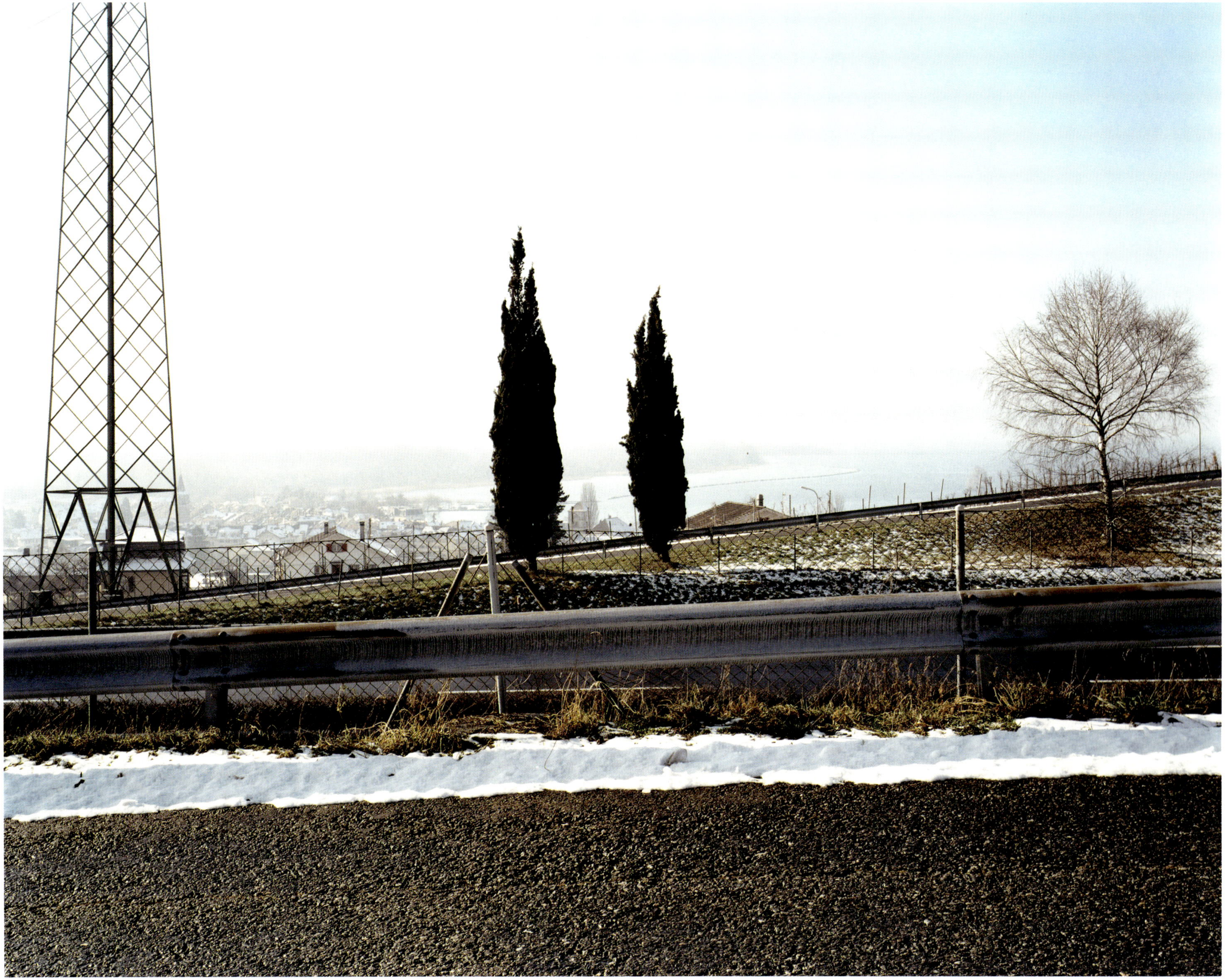

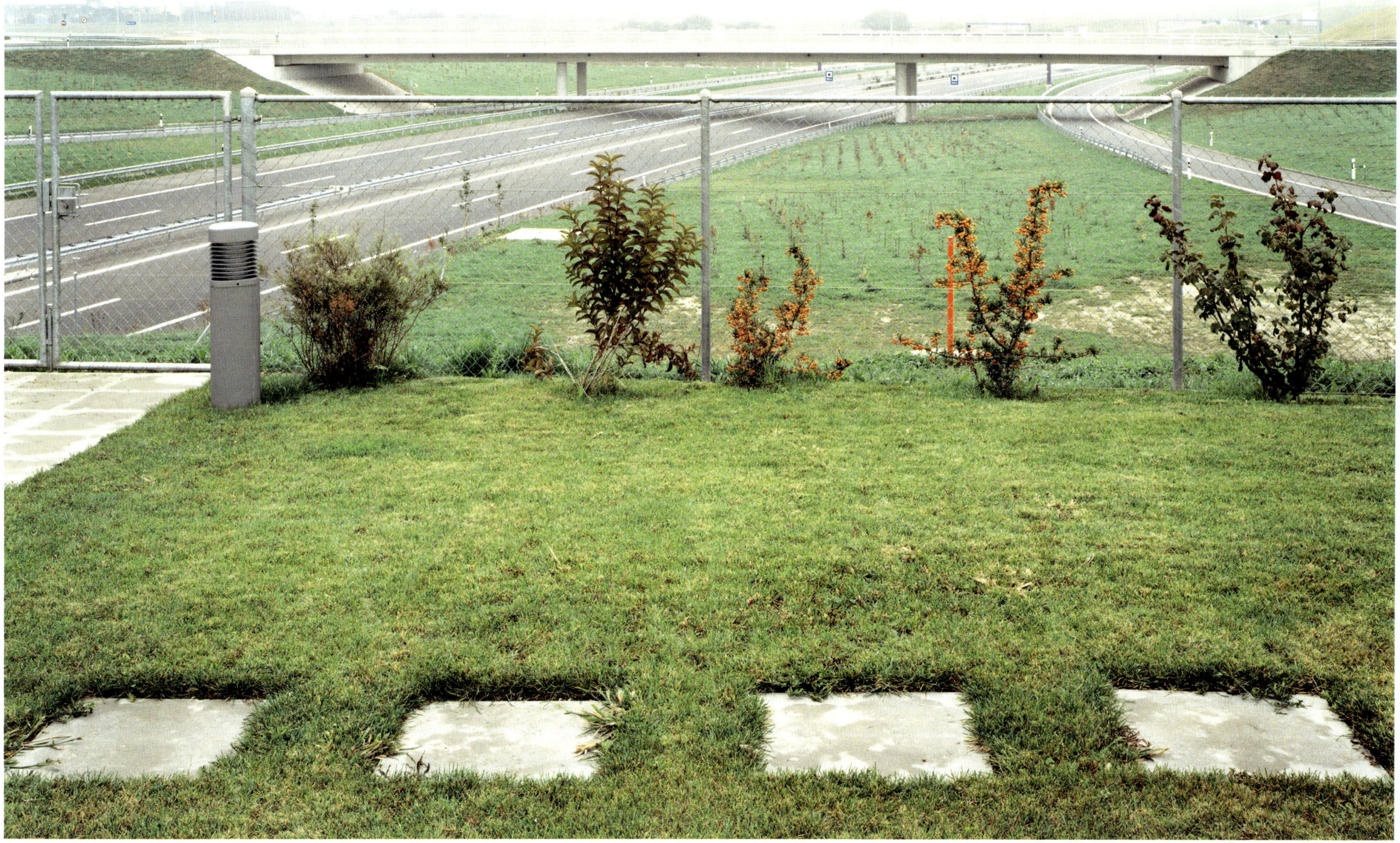

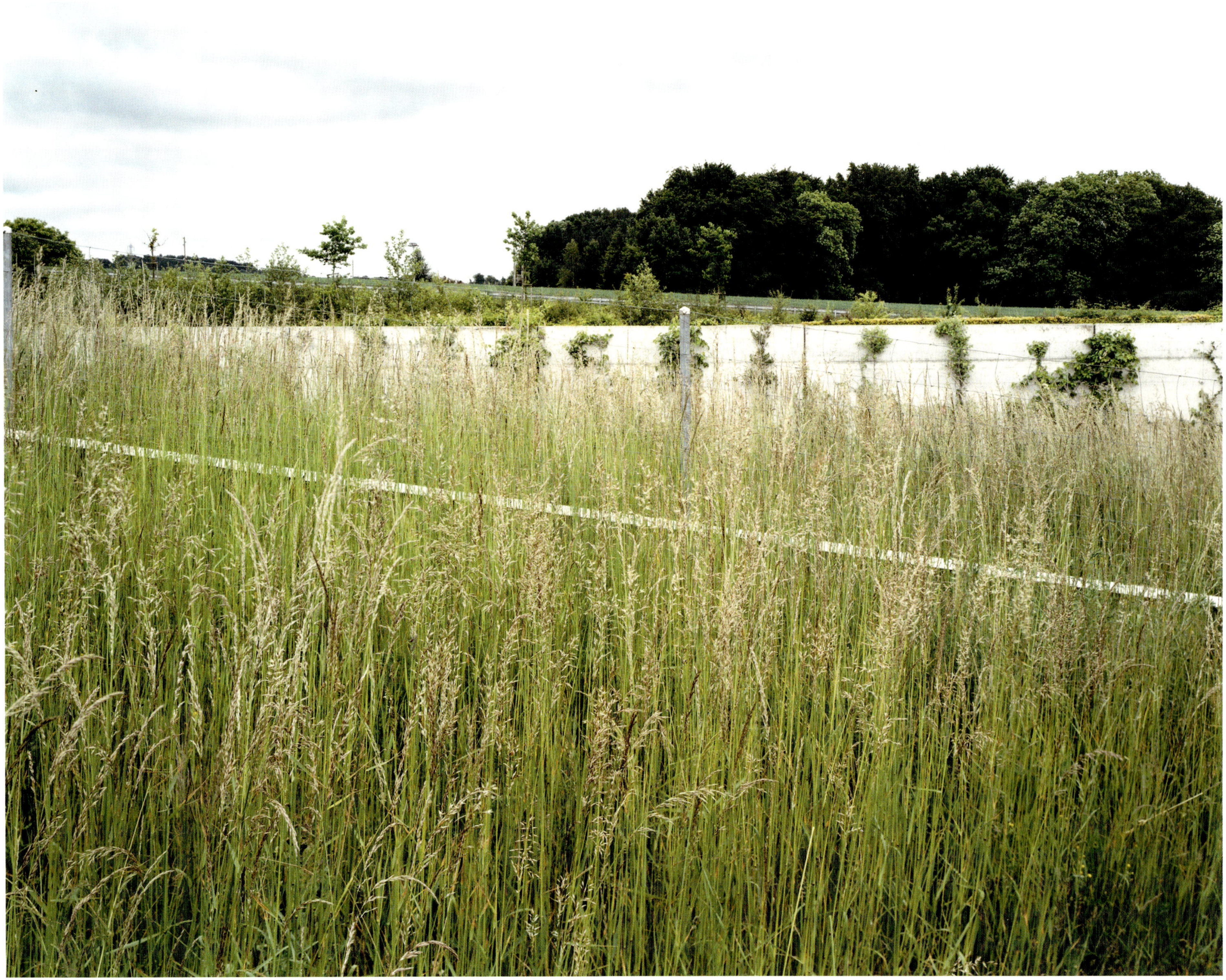

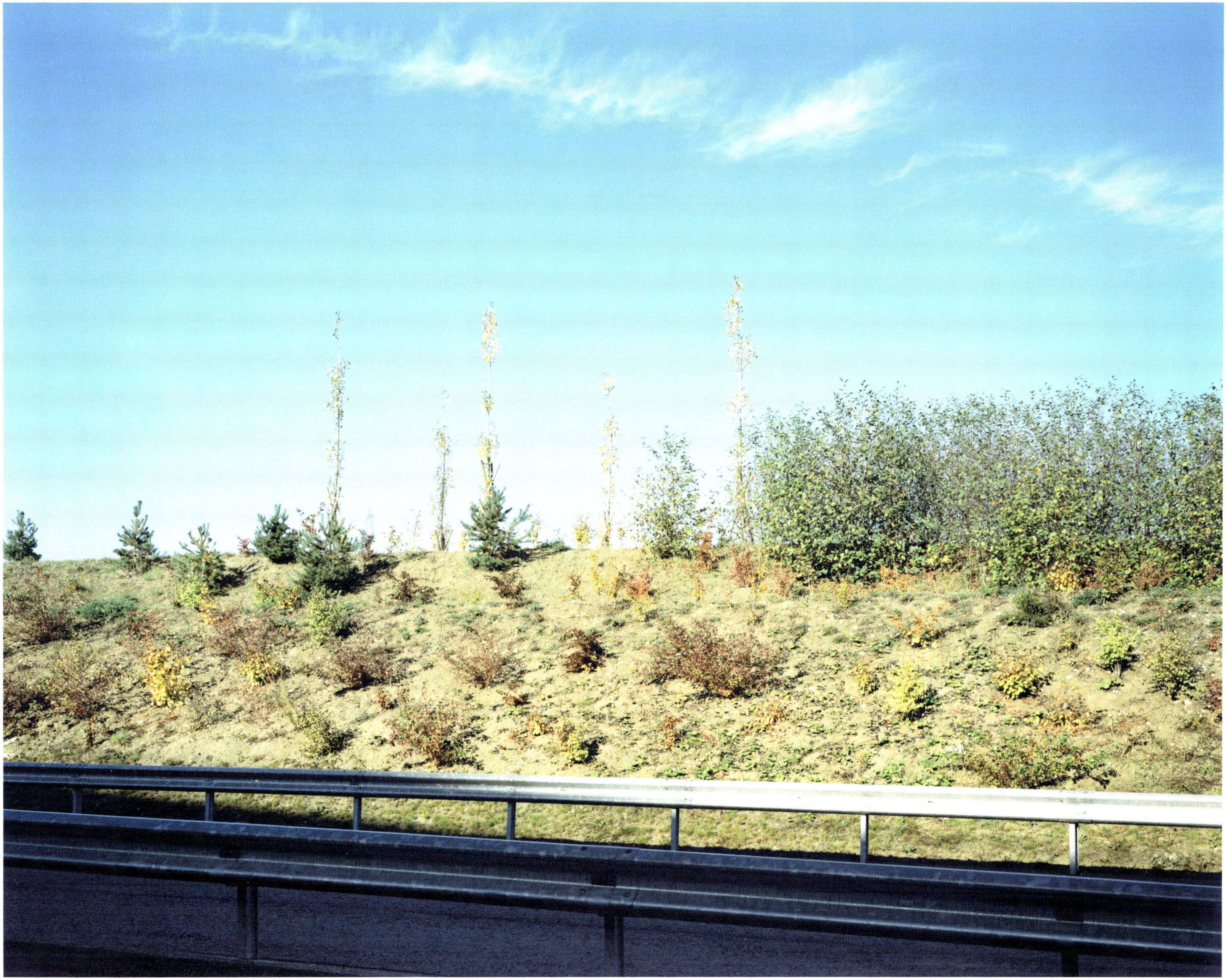

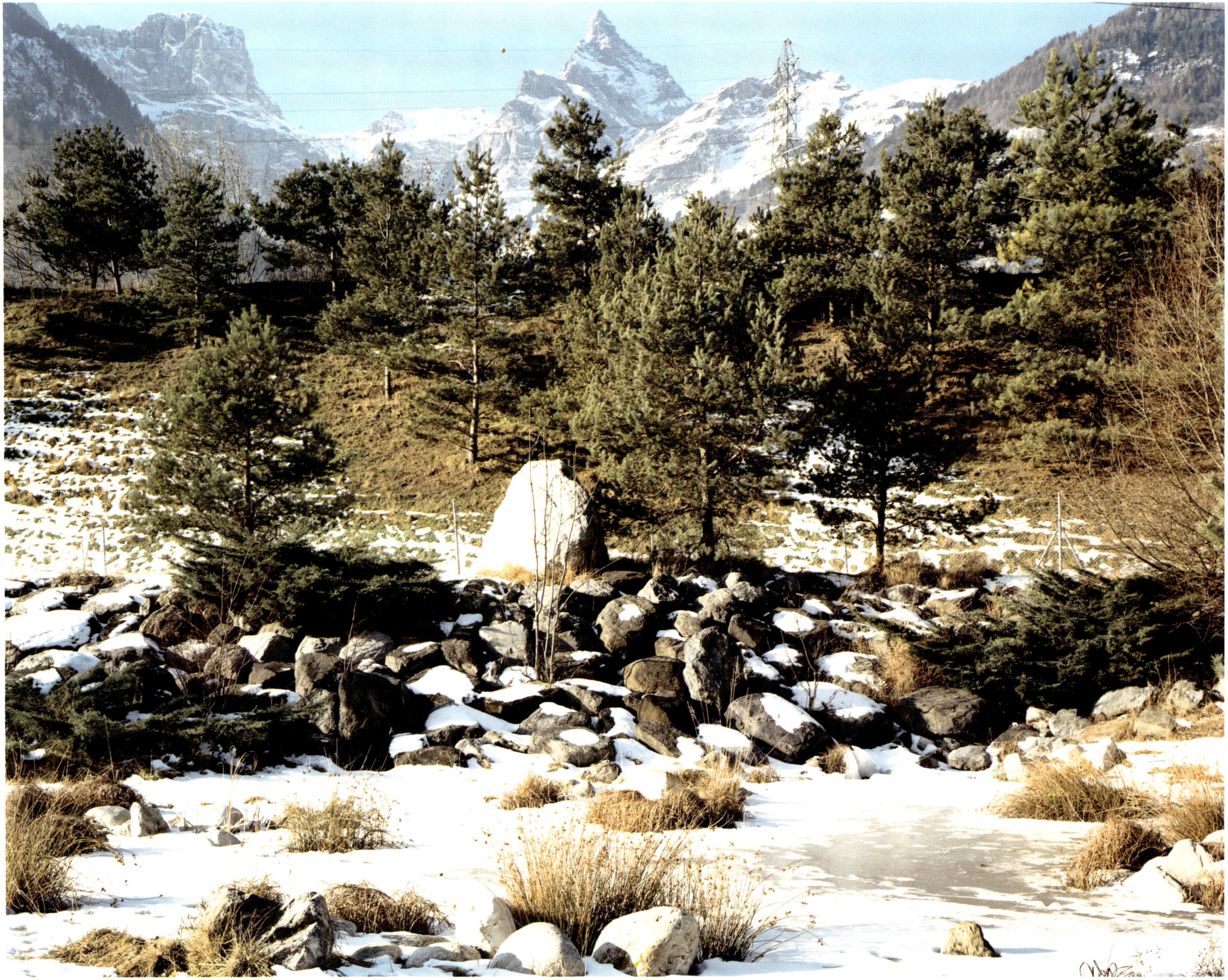

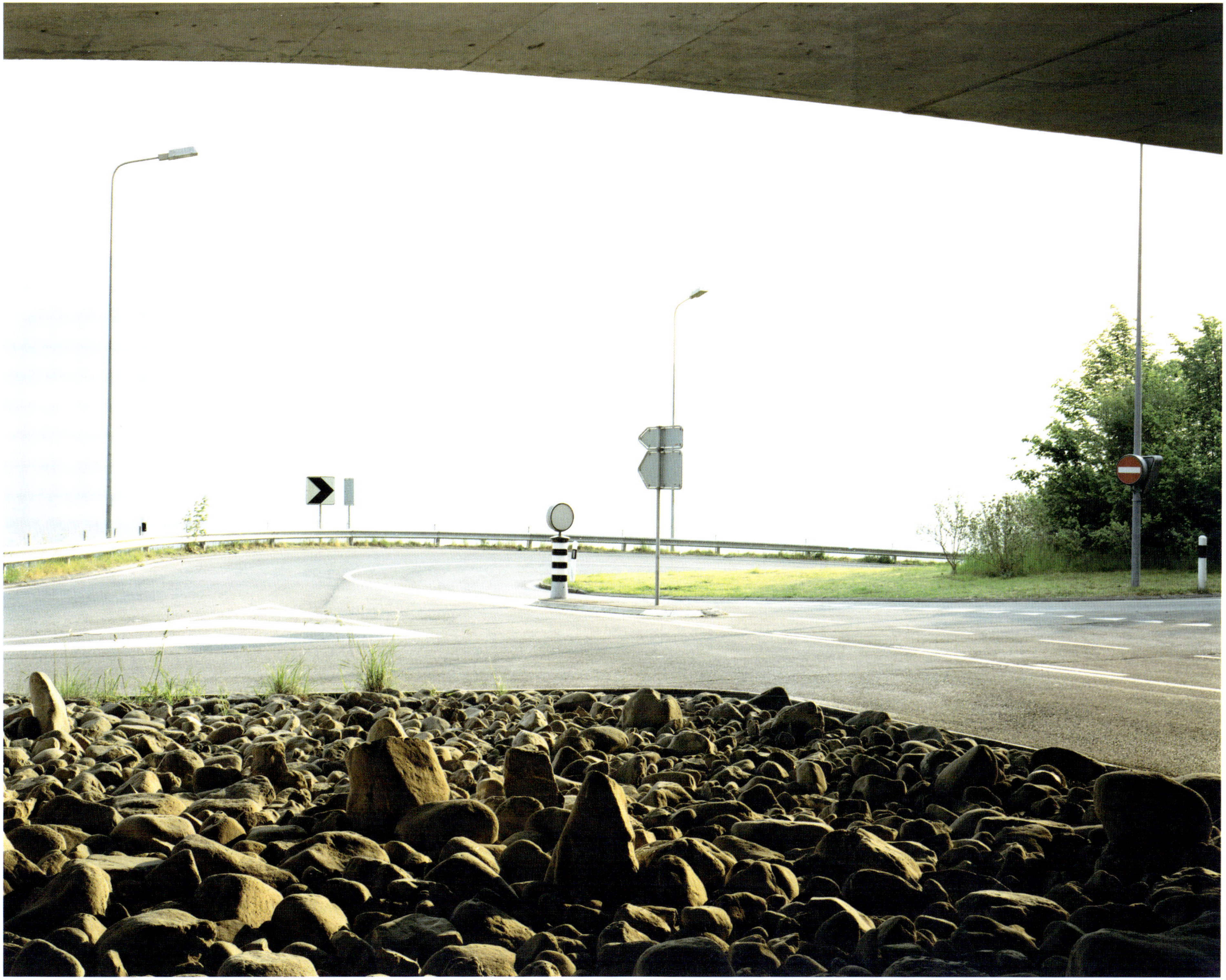

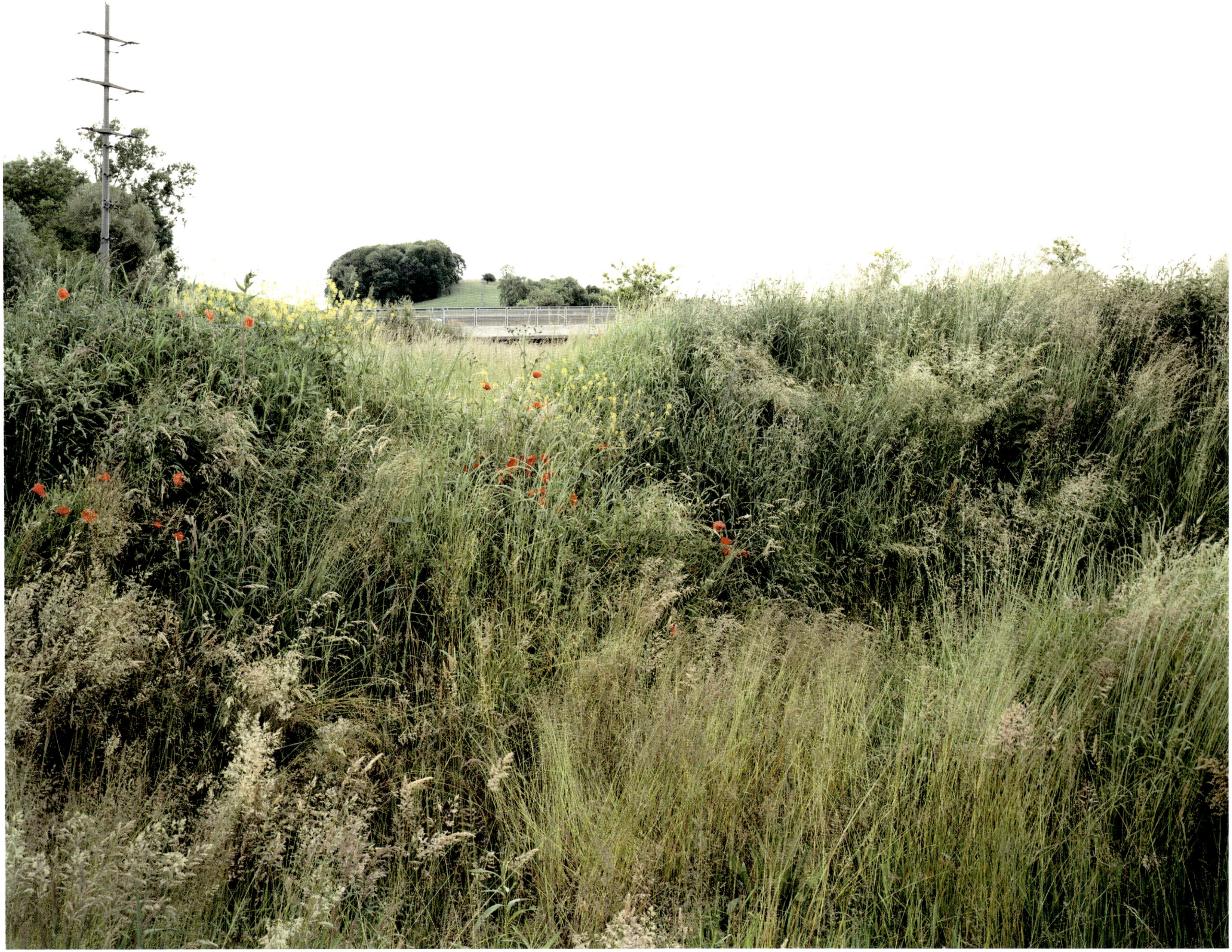

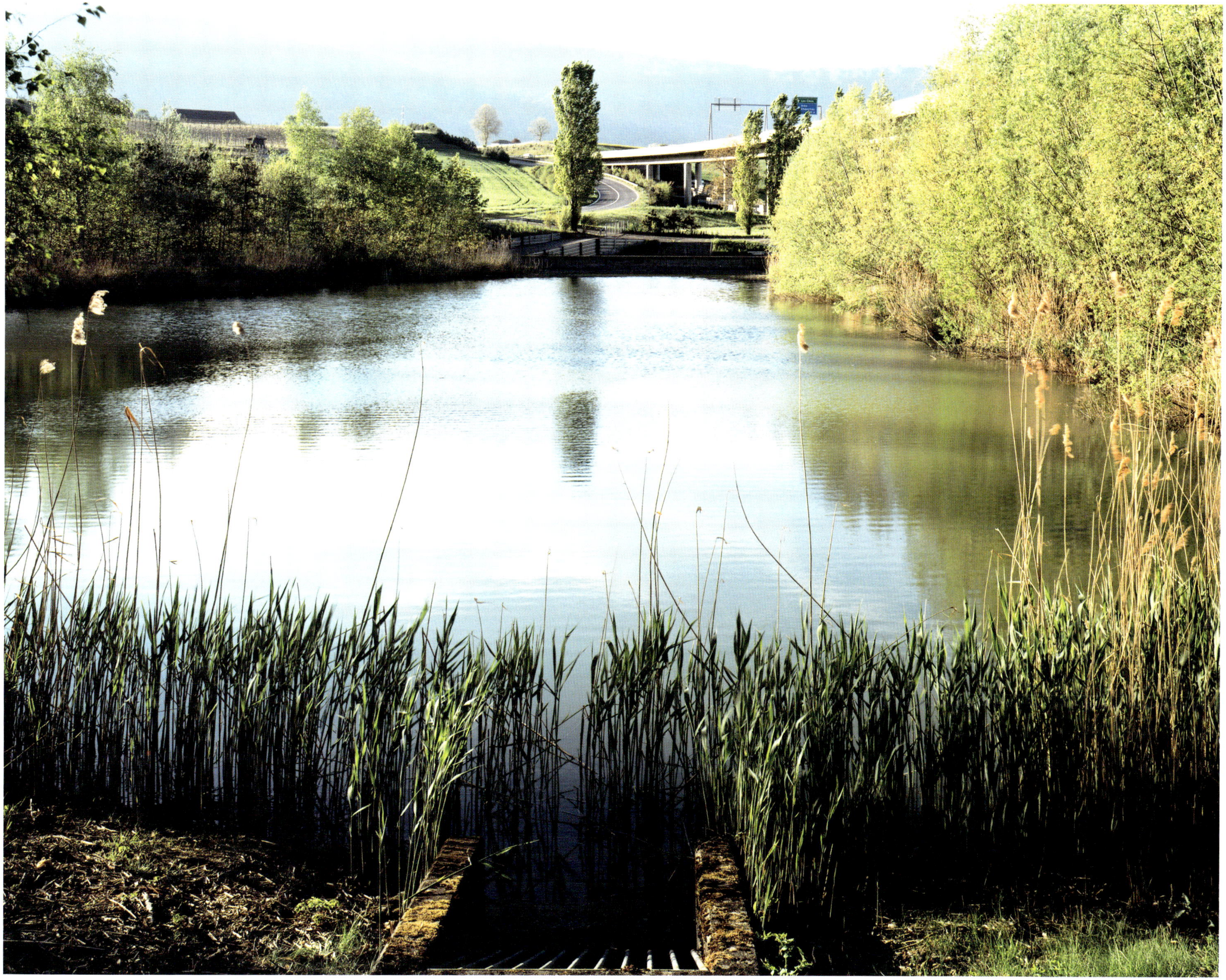

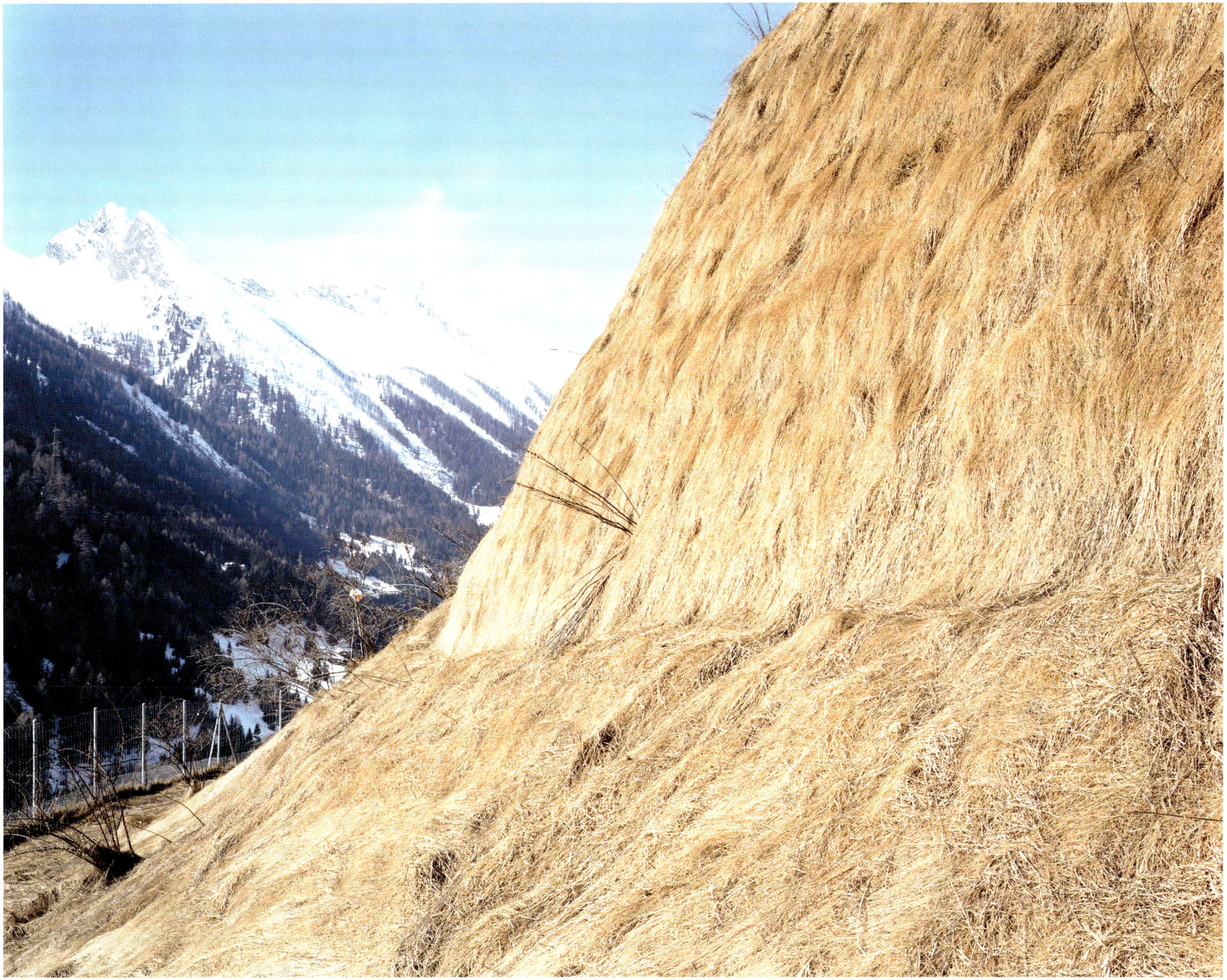

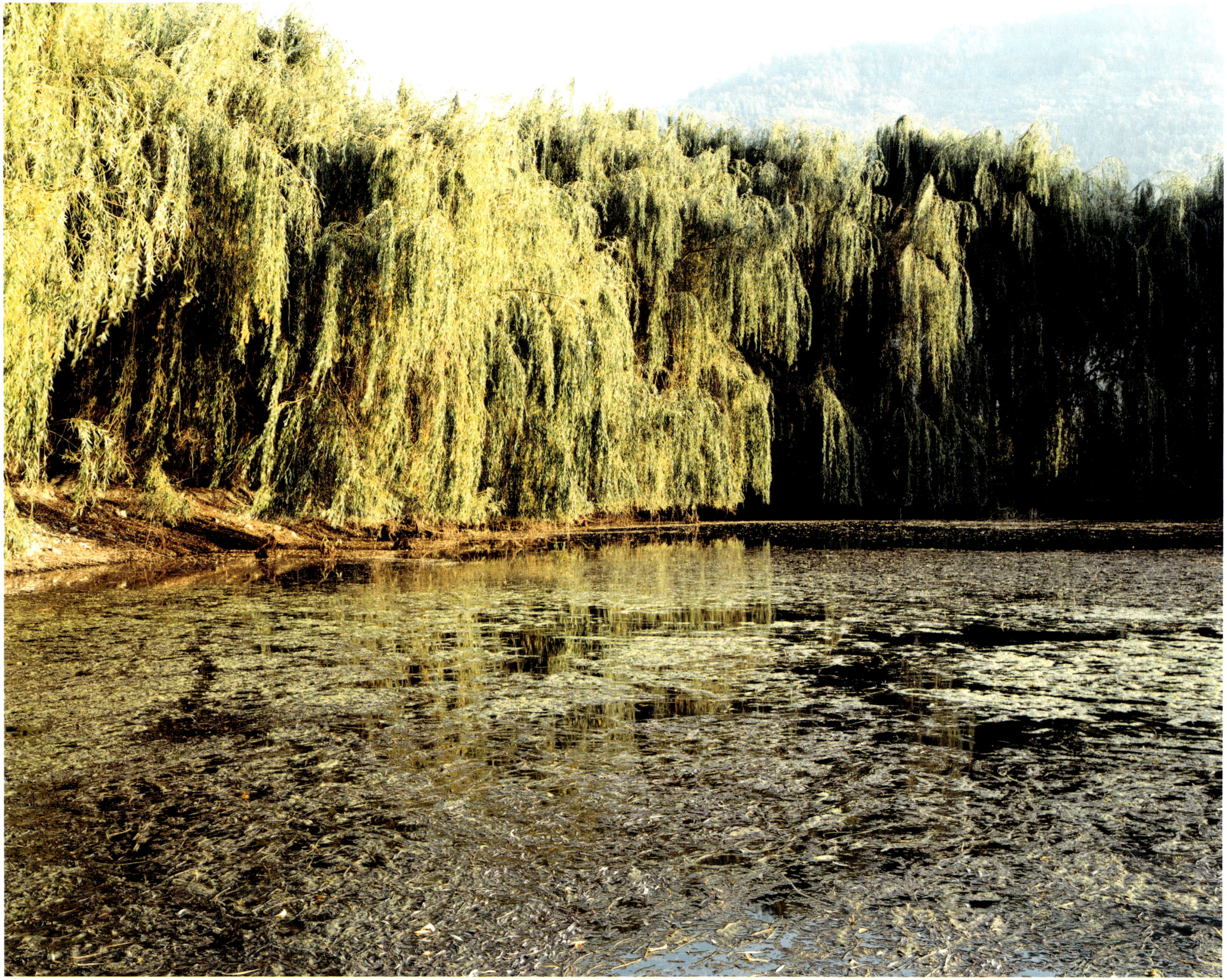

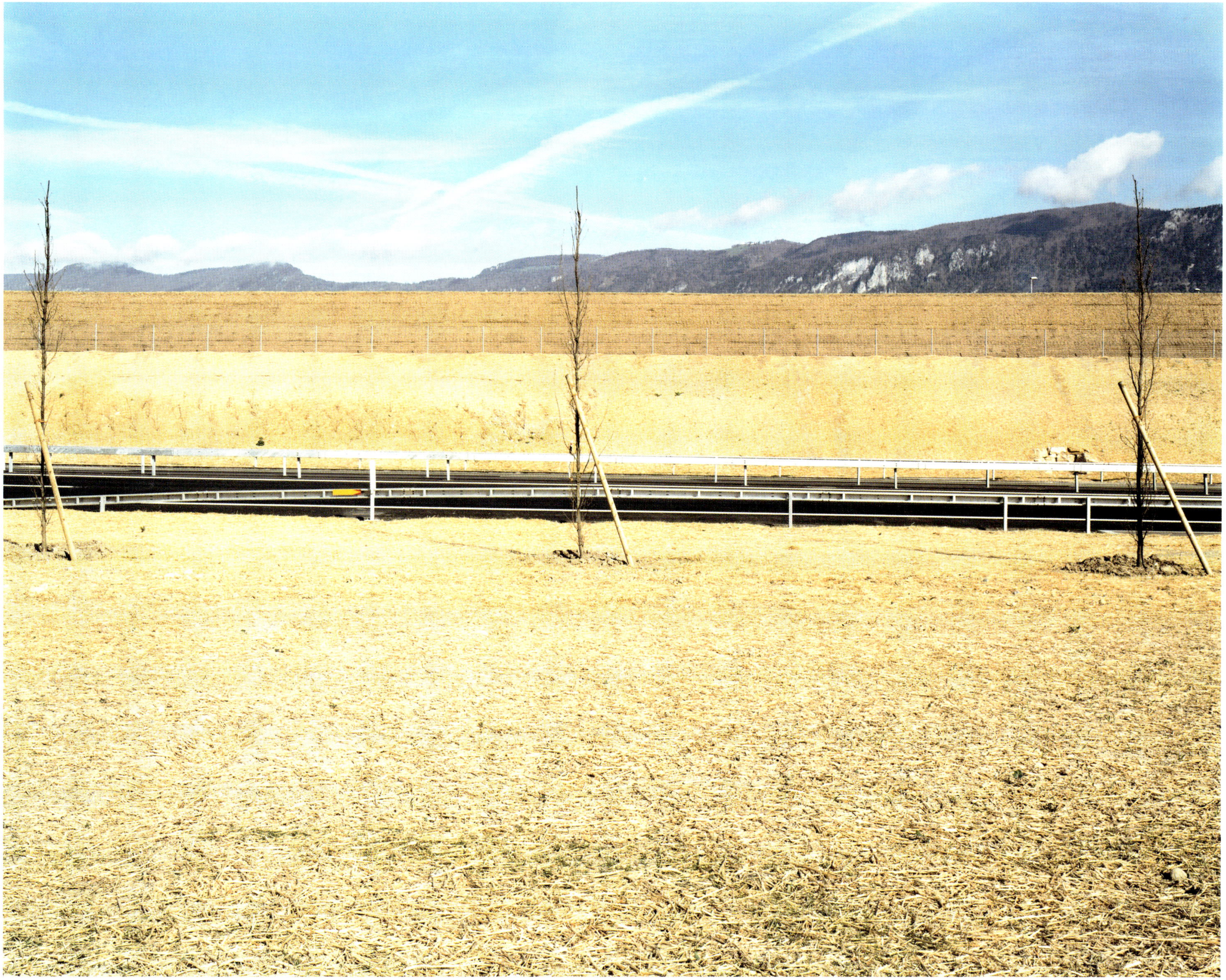

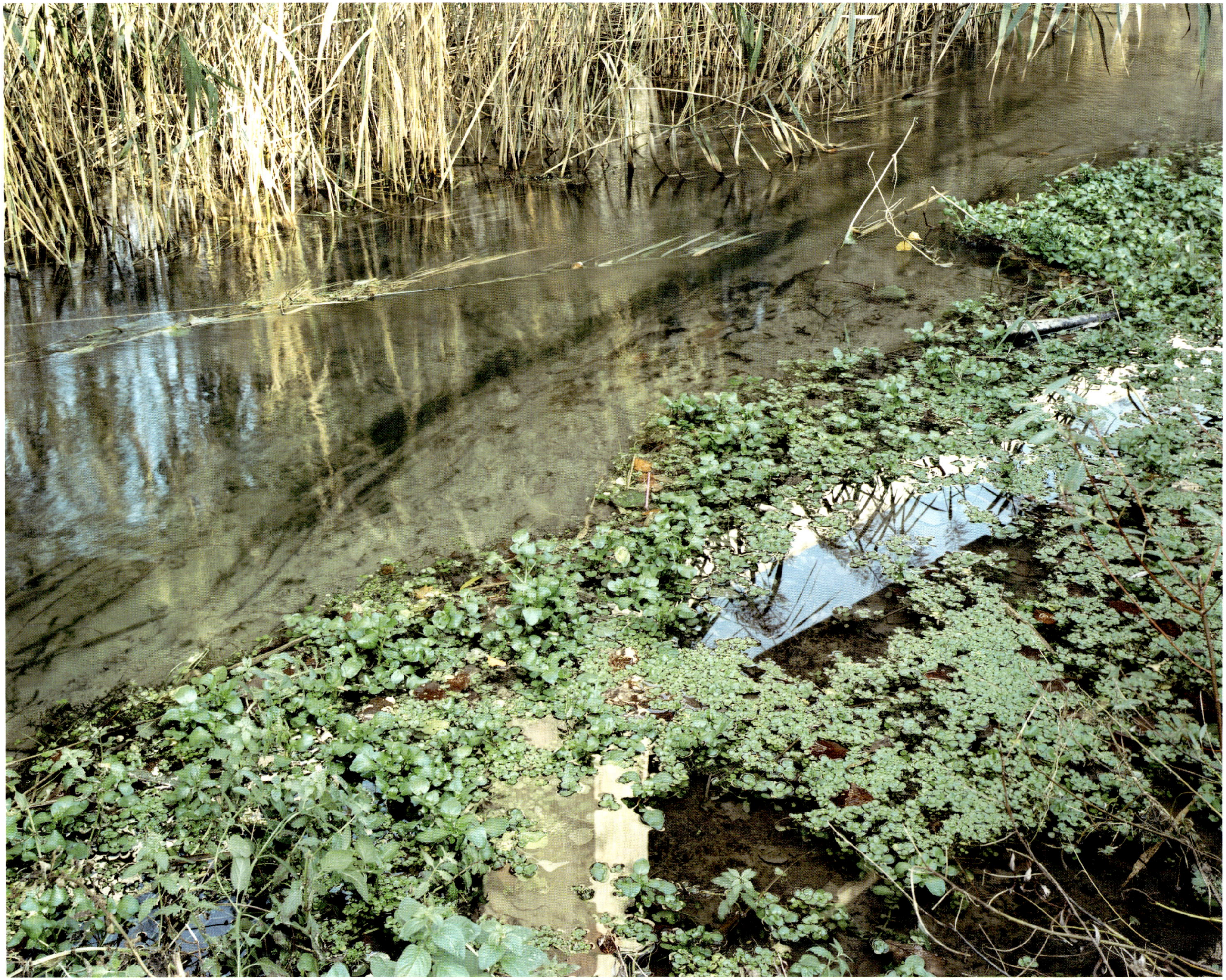

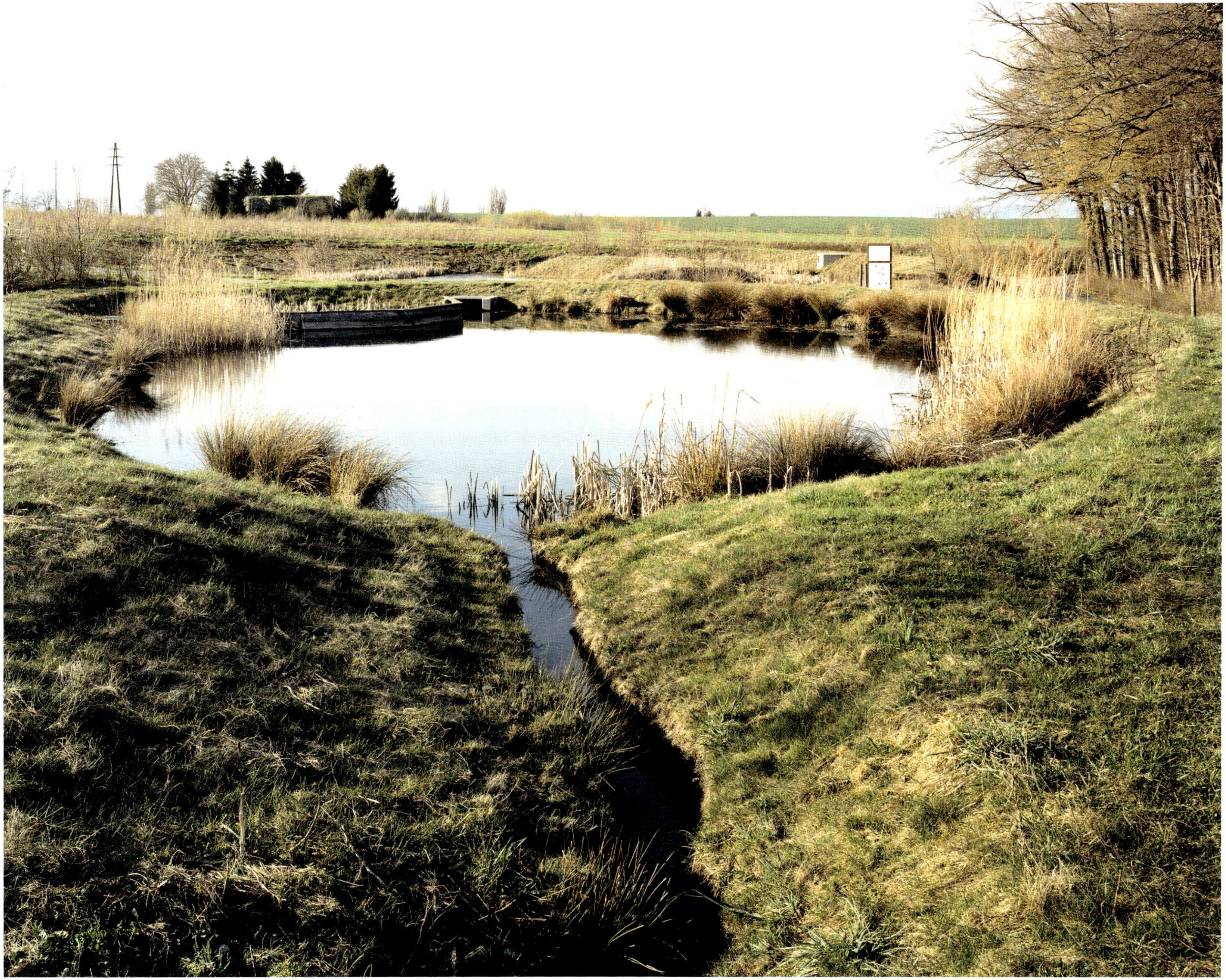

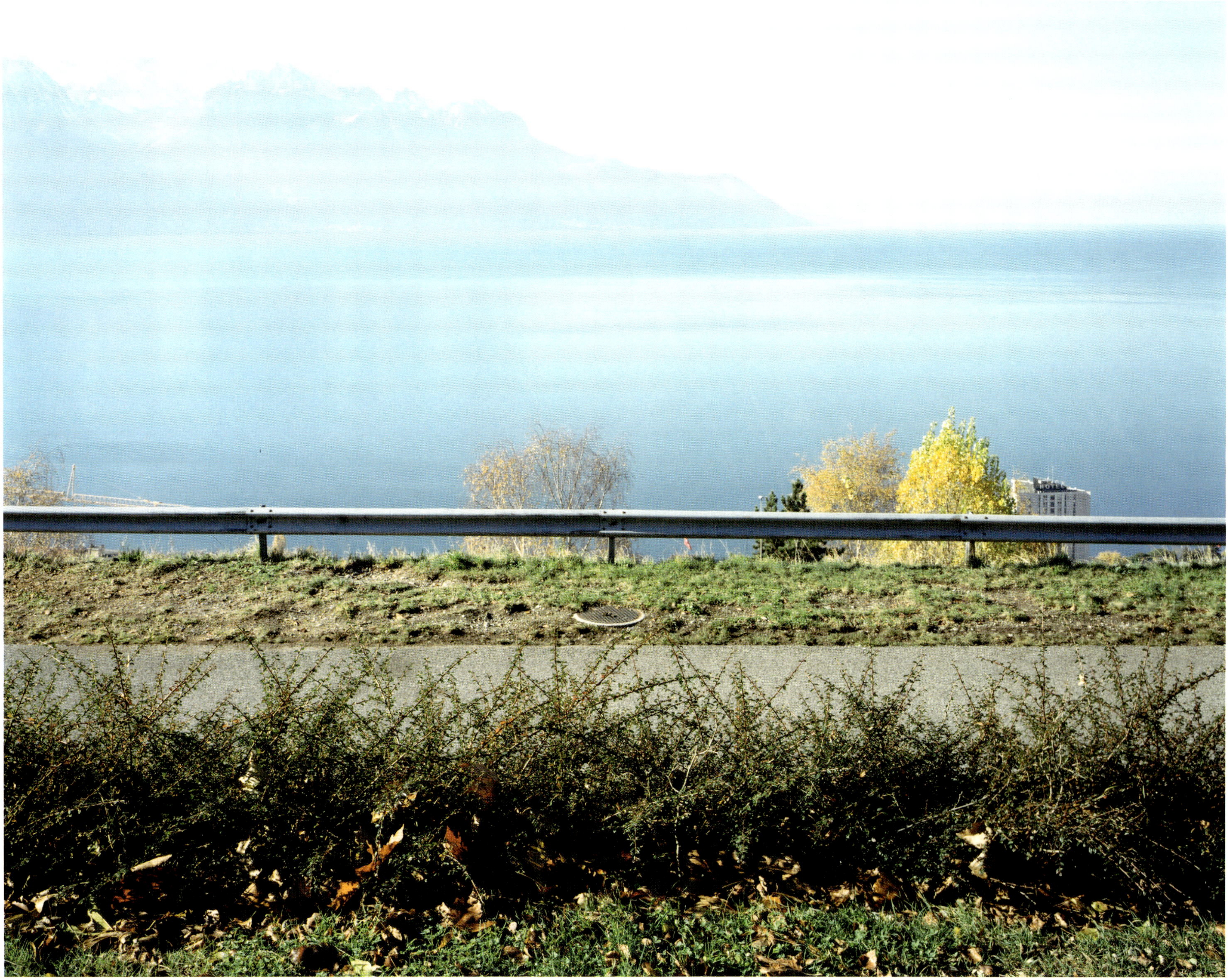

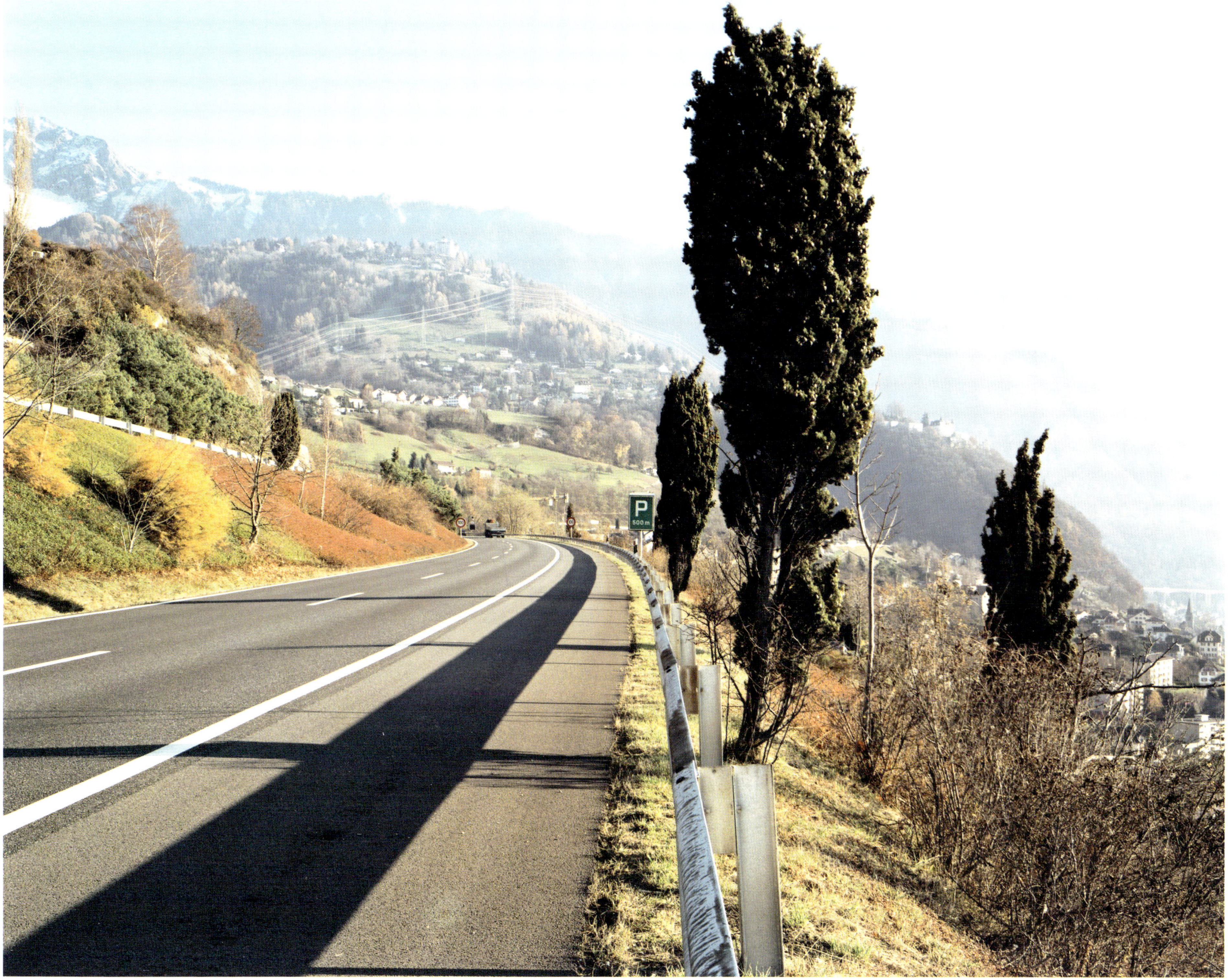
P
500 m

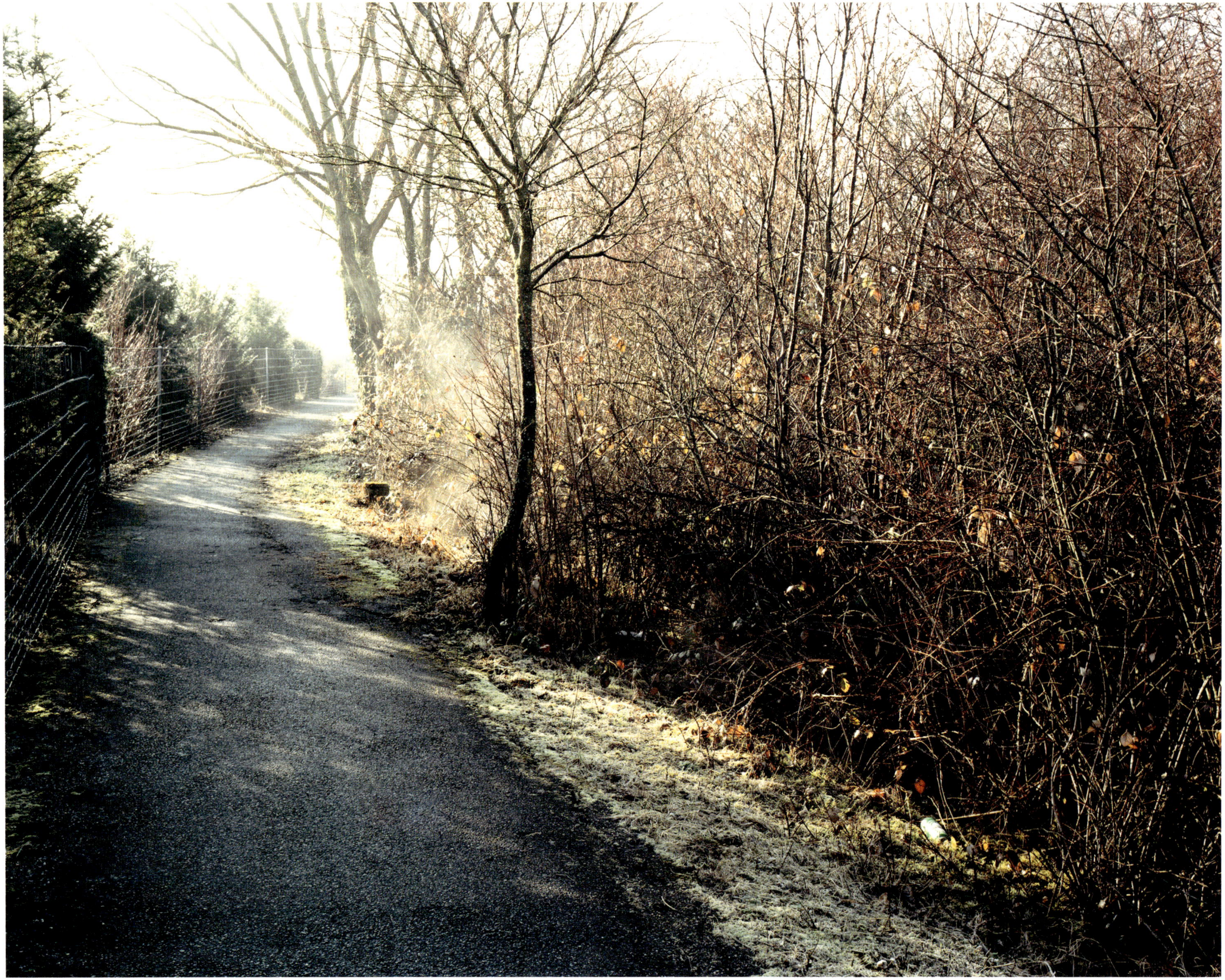

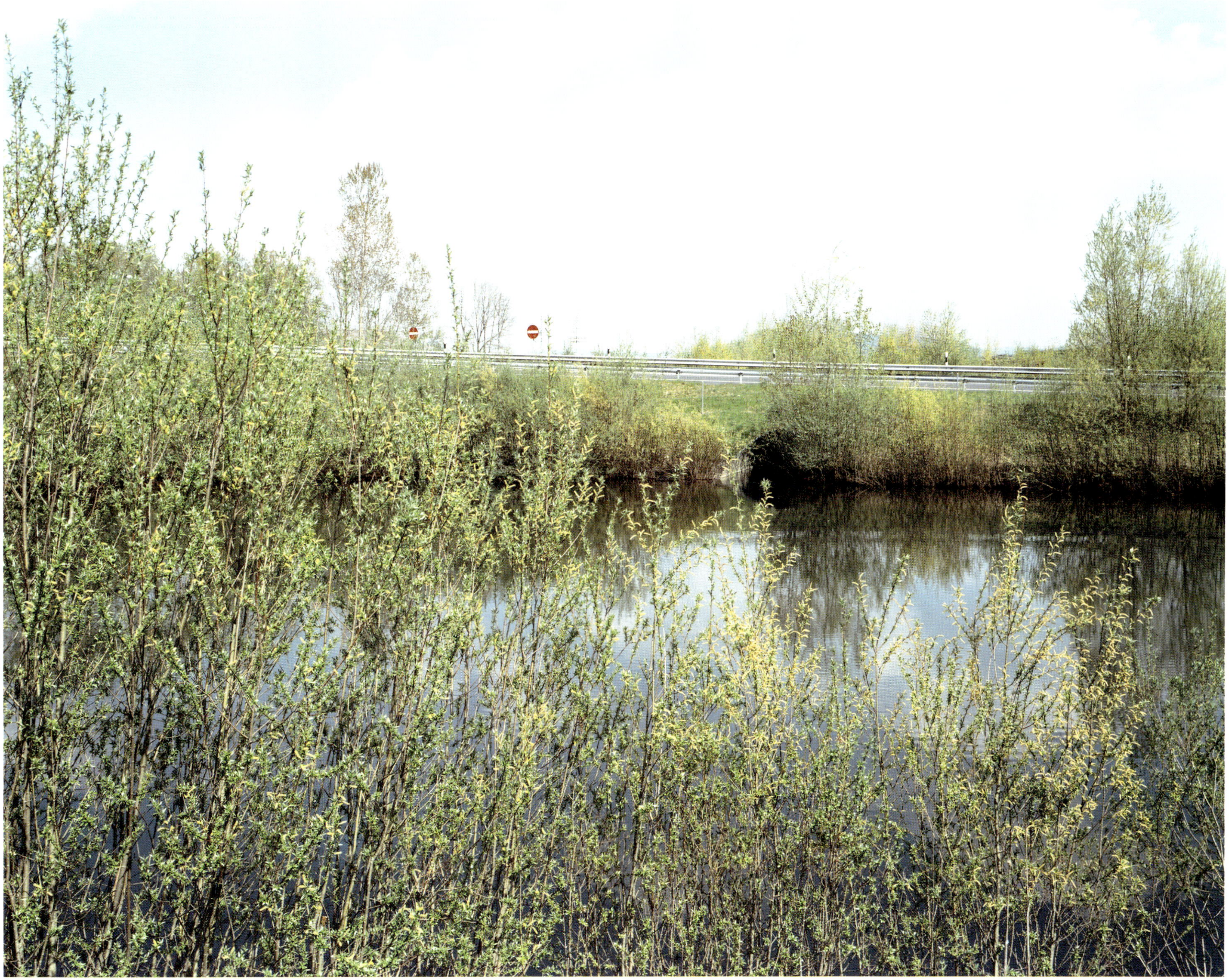

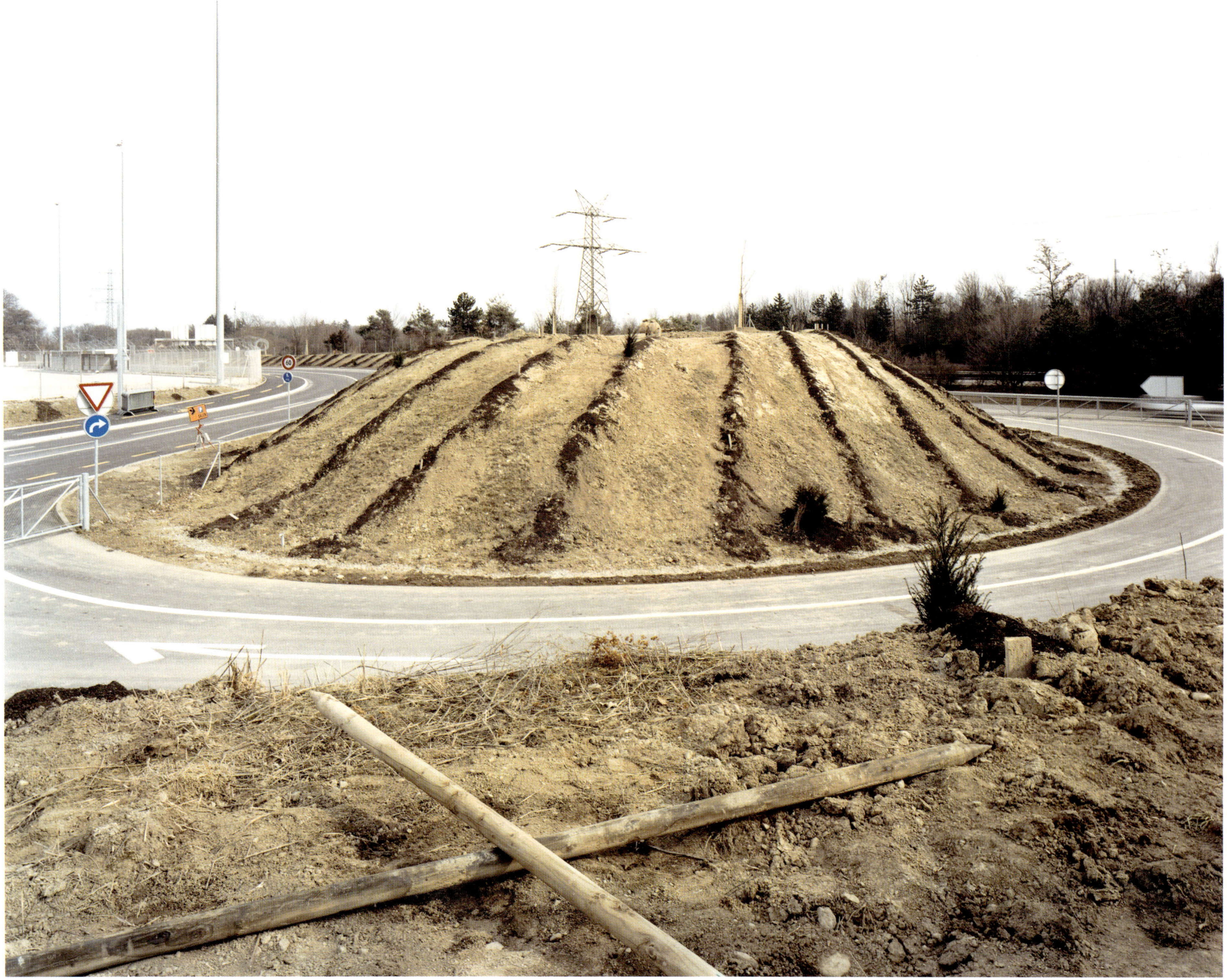
60

Biography and selected exhibitions

1975 Self taught in photography, Geneva
Moves to New York
1977–1982 Photoreporter for the magazines *GEO*, *Time Life Books*, *Actuel*, *L'Illustré*, etc
Since 1991 Teaches photography at the Ecole cantonale d'art of Lausanne (ECAL)

Swiss Design Award 1980, 1986, 1988
Kodak Price 1992

Solo exhibitions

1991 *Pierres fétiches*, Musée de l'Elysée, Lausanne
1998 *Von einer Schweiz zur anderen, De Suisse en Suisse*, Fotomuseum, Winterthur
1999 *Die Schweizer Autobahn*, Museum für Gestaltung, Zurich
2005 *Landscape A*, Musée de l'Elysée, Lausanne

Group exhibitions

1979 *American Vision*, New York
1981 *Jeunes photographes suisses*, Musée d'Art et d'Histoire, Geneva
Swiss Foundation of Photography, Kunsthaus, Zurich
1982 *New Acquisitions*, MoMa, New York
1987 *Arc Lémanique*, Galerie Halle Sud, Geneva / Musée de l'Elysée, Lausanne
1990 *Wichtige Bilder*, Museum für Gestaltung, Zurich
1991 National Foundation of Photographie, Lyon
1992 *Uberall ist Jemand, Räume in besetzen Land*, Museum für Gestaltung, Zurich
Sprung in die Zeit, Berlinische Galerie, Berlin
Fotokunst aus der Schweiz, Forum für zeitgenössische Fotographie, Munich
Biennale, Rotterdam
Swiss Cultural Center, Paris
1993 *Aus der Romandie*, Fotomuseum, Winterthur
1994 The Month of Photography, Bratislava
Les Rencontres photographiques en Bretagne, Lorient
1995 Center of Photography, Geneva
1996 Swiss Foundation of Photography, Kiev
1997 *Voglio vedere le mie montagne, Die Schwerkraft der Berge 1774–1997*, Aargauer Kunsthaus, Aarau
Alpenblick, Kunsthalle Wien
1998 *Fünf Komma Fünf*, Fotomuseum, Winterthur
Architektur im 20. Jahrhundert: Schweiz, Deutsches Architektur-Museum, Frankfort
1999 *Die Schweizer Autobahn*, Museum im Bellpark, Kriens
2000 *CH – Aktuelle Kunst aus der Schweiz*, Ludwig Museum of Contemporary Art, Budapest
Fotohof, Salzburg
Photoforum PasquArt, Biel
Musée d'art moderne et contemporain, Strasbourg
MAMCO, Geneva
2002 *Vivement 2002*, MAMCO, Geneva
De natura rerum, galerie Baudoin Lebon, Paris
Life of the City, MoMa, New York
Help L'elac !, L'elac, Lausanne
Backlight 02. Critical Authenticity, Museum Centre Vapriikki, Tampere
2004 *50-JPG*, Center of Photography / Center for Contemporary Art, Geneva
Interferences, Contemporary Swiss Photography, Center of Photography, Copenhagen

Bibliography

Nicolas Faure, *Goodbye Manhattan*, Panorama Verlag, Altstätten, 1984
Nicolas Faure, *Switzerland on the rocks*, Scalo and Museum für Gestaltung, Zürich – Berlin – New York, 1992
Nicolas Faure and André Klopmann, *Citoyens du monde – Meyrin*, Scalo et Meyrin, 1995
Nicolas Faure, *Von einer Schweiz zur anderen / De Suisse en Suisse*, Fotomuseum Winterthur and Scalo, Zürich – Berlin – New York, 1998
Nicolas Faure, *Autoland, Bilder aus der Schweiz / Pictures from Switzerland*, Museum für Gestaltung Zürich and Scalo, Zürich – Berlin – New York, 1999
Nicolas Faure, Philippe Lacoue-Labarthe, Jean-Luc Nancy, *Portraits / Chantiers*, Mamco, Geneva, 2004

Books from group exhibitions

Jeunes photographes suisses, Fondation pour la photographie, Musée d'Art et d'Histoire, Geneva, 1981
Arc Lémanique, 21 récits sur le lieu, P. M. Favre et L'Hebdo, Lausanne, 1988
Wichtige Bilder, Fotografie in der Schweiz, Urs Stahel & Martin Heller, Der Altag and Museum für Gestaltung, Zürich, 1990
Voir la Suisse autrement / Die Schweiz anders sehen / Un altro sguardo sulla Svizzera / Vesair la Svizra en moda diversa, Musée d'Art et d'Histoire, Fribourg, 1991
Nouveaux itinéraires / Neue Wege / Nuovi itinerari / Novs trajects. Lausanne, Musée de l'Elysée, 1991
Aus der Romandie, Fotomuseum Winterthur, 1993

First edition 2005

Steidl
Düstere Strasse 4
D–37073 Göttingen
T +49 551 49 60 60
F +49 551 49 60 649
mail@steidl.de
www.steidlville.com
www.steidl.de

Musée de l'Elysée
18, avenue de l'Elysée
CH–1014 Lausanne
T +41 (0)21 316 99 11
F +41 (0)21 316 99 12
musee.elysee@serac.vd.ch

ISBN 3–86521–212–3
Printed in Germany

Graphic design
NORM, Dimitri Bruni & Manuel Krebs, Zurich

Printing
Steidl, Göttingen

Set in
Theinhardt-Light

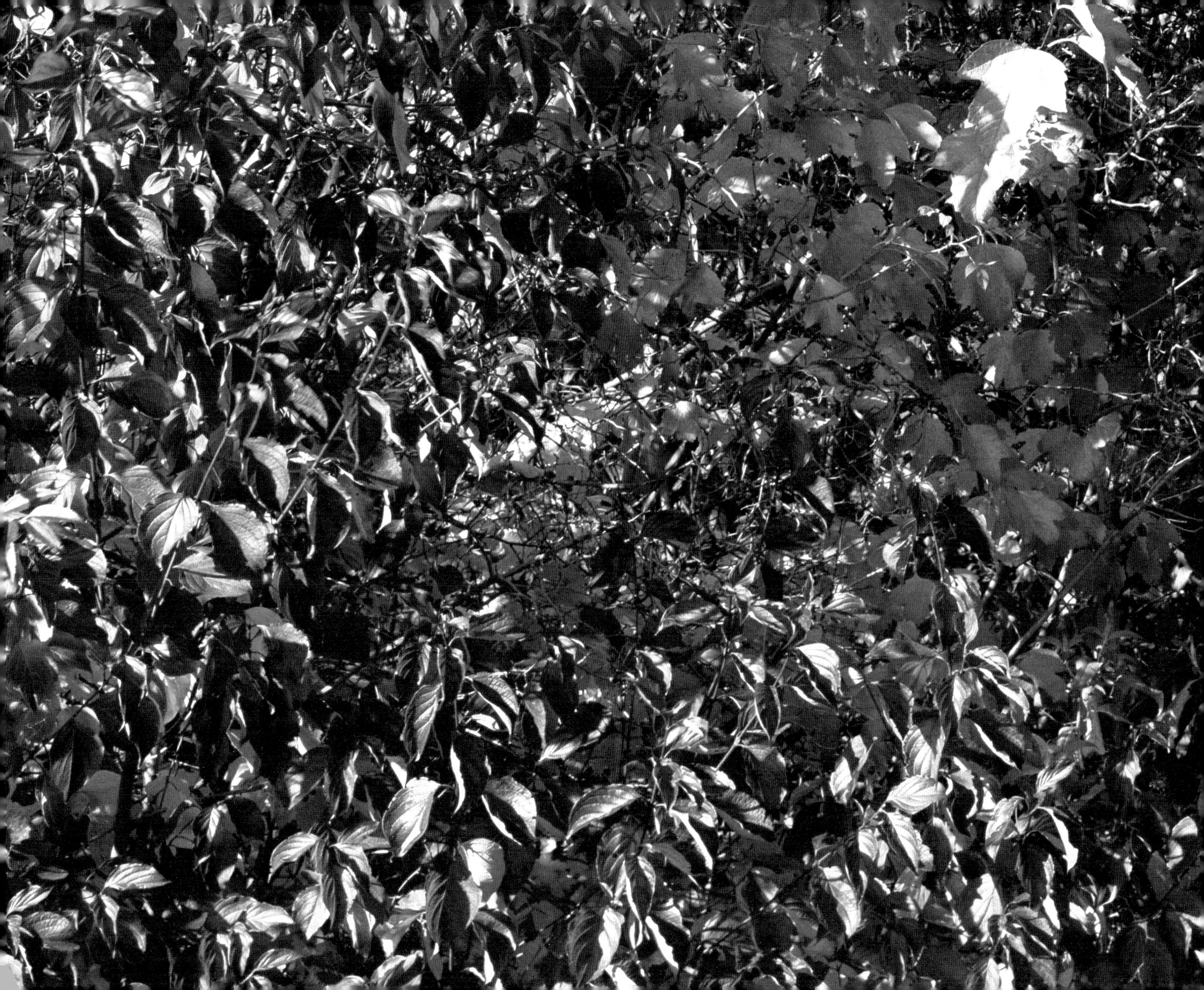

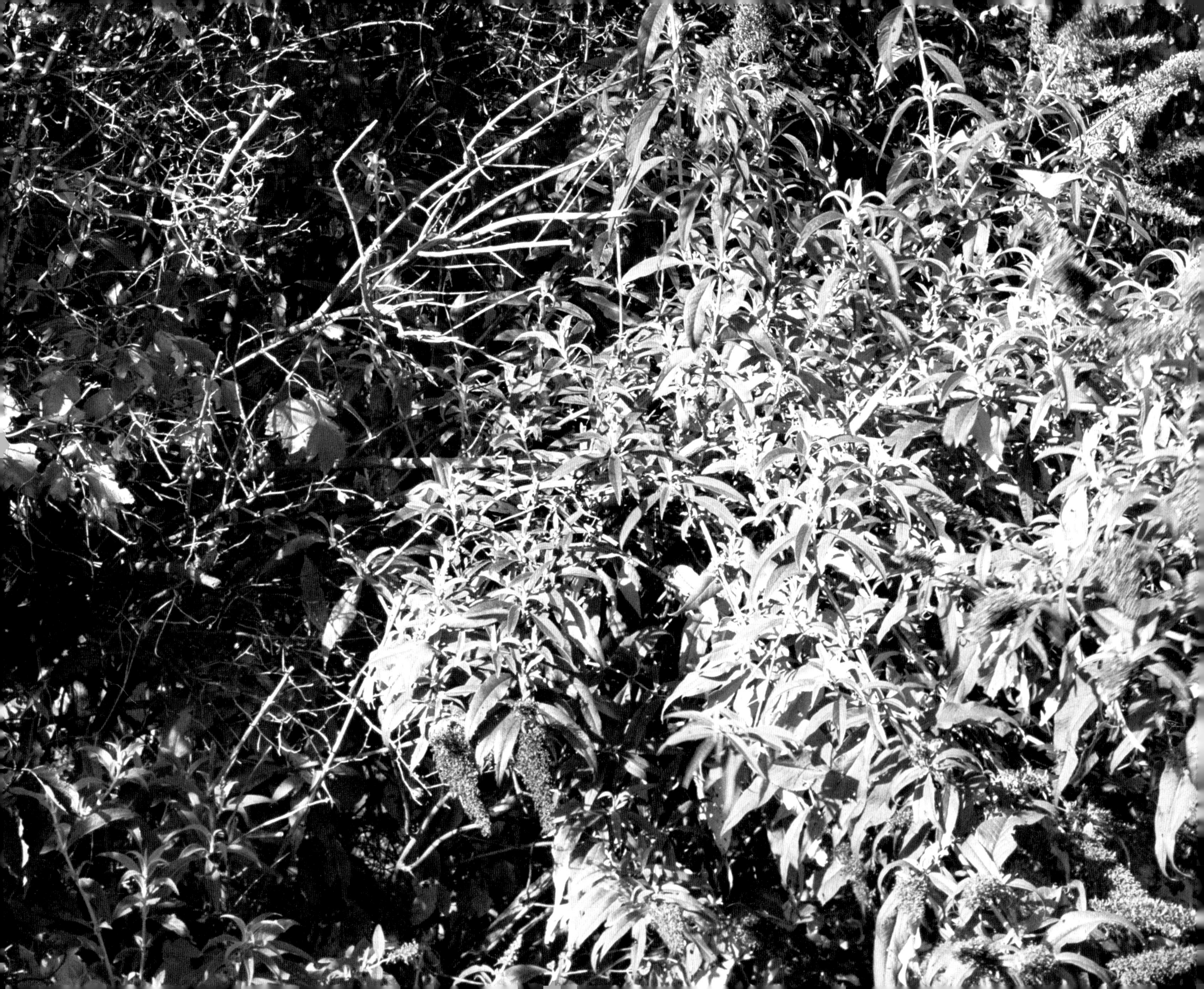